BASICS
PRODUCT DESIGN

David Bramston
and YeLi

Idea Searching for Design

How to Research and Develop Design Concepts

Second Edition

BLOOMSBURY VISUAL ARTS
LONDON · NEW YORK · OXFORD · NEW DELHI · SYDNEY

BLOOMSBURY VISUAL ARTS
Bloomsbury Publishing Plc
50 Bedford Square, London, WC1B 3DP, UK
1385 Broadway, New York, NY 10018, USA
29 Earlsfort Terrace, Dublin 2, Ireland

BLOOMSBURY, BLOOMSBURY VISUAL ARTS and the
Diana logo are trademarks of Bloomsbury Publishing Plc

First published in Great Britain by AVA Publishing SA 2009
Second edition published by Fairchild Books 2016
This edition published by Bloomsbury Visual Arts 2019
Reprinted 2020, 2022, 2023

Copyright © Bloomsbury Publishing Plc, 2016

David Bramston and YeLi have asserted their rights
under the Copyright, Designs and Patents Act, 1988,
to be identified as Author of this work.

For legal purposes the Acknowledgements on p. 184
constitute an extension of this copyright page.

All rights reserved. No part of this publication may be
reproduced or transmitted in any form or by any means,
electronic or mechanical, including photocopying,
recording, or any information storage or retrieval system,
without prior permission in writing from the publishers.

Bloomsbury Publishing Plc does not have any control over,
or responsibility for, any third-party websites referred to
or in this book. All internet addresses given in this book
were correct at the time of going to press. The author and
publisher regret any inconvenience caused if addresses
have changed or sites have ceased to exist, but can
accept no responsibility for any such changes.

A catalogue record for this book is available from the
British Library.

The Library of Congress has cataloged the Fairchild Books
edition as follows:
Names: Bramston, Dave, author.
Title: Idea searching for design : how to research
    and develop design concepts / by David
    Bramston and YeLi.
Description: Second edition. | New York : Fairchild
    Books, 2016. | Series: Basics product design |
    Includes bibliographical references and index.
Identifiers: LCCN 2015044116| ISBN
    9781472581969 (paperback) | ISBN
    9781472581976 (epdf)
Subjects: LCSH: Product design. | Creative thinking.
    | BISAC: DESIGN / Product. | TECHNOLOGY &
    ENGINEERING / Industrial Design / Product.
Classification: LCC TS171 .B695 2016 | DDC
    658.5/752--dc23 LC record available at http://
    lccn.loc.gov/2015044116

ISBN: PB: 978-1-3501-4079-0
       ePDF: 978-1-4725-8197-6

Series: Basics Product Design

Typeset by Saxon Graphics Ltd, Derby
Printed and bound in Great Britain

To find out more about our authors and books visit
www.bloomsbury.com and sign up for our newsletters.

0.1

0.1
Matthias Pliessnig
Platinum (Beirut Lebanon)
Photo credit: Ieva Saudargaitė

Contents 4

Introduction 6

Chapter 1
Just Imagine if it were Possible 8

Absorb 10

Thoughts 19

Observations 42

Chapter 2
Thinking Differently 62

Understanding 64

Profiles 74

Themes 75

Chapter 3
Experimental Beauty 86

Materials 88

Exploration 92

Communication 102

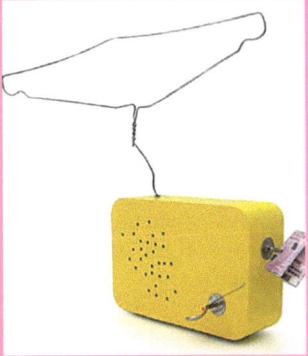

Chapter 4 **Sensory** **Perceptions**	**104**
Sensory	106
Added values	115
Conflicts	123

Chapter 5 **Adopting** **Responsibility**	**132**
Sustainability	134
Tinkering	138
Inspirational	148

Chapter 6 **Evolving** **the Reality**	**150**
Blue sky	152

Project briefs	156
Conclusion	169
Glossary	170
Bibliography	173
Contacts	176
Index	177
Credits	180
Acknowledgments	184

Introduction

Diverse conscious and subconscious experiences provide a unique, ongoing mental catalogue of references capable of assisting the generation of ideas if encouraged to do so. A continual journey of experiencing the familiar along with the unfamiliar naturally broadens the imagination.

All experiences are viable encounters for this purpose, and often it is those that at first appear to be too abstract or irrational which become the essential catalyst for the most interesting suggestions and directions in the end.

Everyone has their own set of experiences, and an opportunity to engage in dialogue with others who are both within and beyond one's immediate circle of associates encourages creative variance and can further stimulate authentic thinking. In the initial stages of idea generation everyone is different, and everyone has a different take on things due to the exclusivity of their personal backgrounds.

The ability to engage the senses as critical tools and to attempt to see things in an alternative context rather than being mentored or escorted by any inherent mental baggage and preconceived values encourages original thought. The French novelist and social commentator Marcel Proust (1871–1922) recognised the importance of "seeing with new eyes," a view echoed by Henry David Thoreau (1817–1862), who commented that "it is not what you look at that matters it is what you see."

The ability to see what others overlook is a trait of many successful designers. David Kelley at IDEO believes it is possible to see something original in a situation that has perhaps become too familiar, and it is this capacity to discover and take inspiration from everywhere that is so important. Things do talk if the observer is prepared to listen.

The Mitate lamps designed by Studio Wieki Somers (2013) combine culture and objects in arrangements that communicate multiple messages to the observer. This reinterpretation of meaning manages to challenge conventional perceptions.

Taking a risk and having a "dare to be different" attitude is necessary if boundaries are to be pushed. The creative designer must aim to occupy the unfamiliar territory that is beyond their usual remit.

The ideas and outputs of designers such as Yvonne Fehling and Jennie Peiz, with works such as Still Lives or Stuhlhockerbank, are distinctive, experimental, and intriguing. The blurring of conventional and unconventional practices and an ability to think differently with a curious mind redefines the boundaries and questions values. Approaching a problem from an alternative or lateral direction, while remaining in control, assists in the attempt to see things from a removed standpoint.

The Eigruob lamp for Kartell designed by the Japanese studio Nendo (2014) concentrates on the space that surrounds the designed object, creating a void in the form of the original Bourgie lamp by Ferruccio Laviani. Exploring the unexpected is both challenging and revealing. Ideas do not have to be overly complicated to be alluring and wonderful. It is important to try and keep the thinking process simple.

Interdisciplinary and cross-disciplinary approaches provide an insight to alternative means, methods, and ideas. The cross-pollination of ideas between creative disciplines or seemingly unrelated disciplines triggers the imagination. It is not always essential to understand what is being observed initially, as it may still prompt thoughts and opportunities. Being outside a comfort zone introduces original cultures and experiences. The exploratory, challenging, and incredible works of designers such as Jannis Huelsen and the Xylinum stool can introduce unexpected elements and inspire atypical thinking.

Idea generation should not be an arduous journey but rather an exploratory path, a pleasurable adventure, where the failure and collapse of a thought is recognized and ultimately considered as contributing to a success. The confidence to make a mistake is a valuable characteristic of many experimenters in the search for a viable outcome and was recognized by Danish designer Verner Panton (1926–1998), who stated that "a failed experiment can be more important than a trivial design," a sentiment that echoed French sculptor Rodin, who originally stated "nothing is a waste of time if you use the experience wisely."

Idea Searching for Design has engaged international artists, designers, and educators to explore a broad array of methods and practices associated with the practice of idea generation. The process is often eclectic and sometimes visceral, where ephemeral practices continually inspire and steer thought and process.

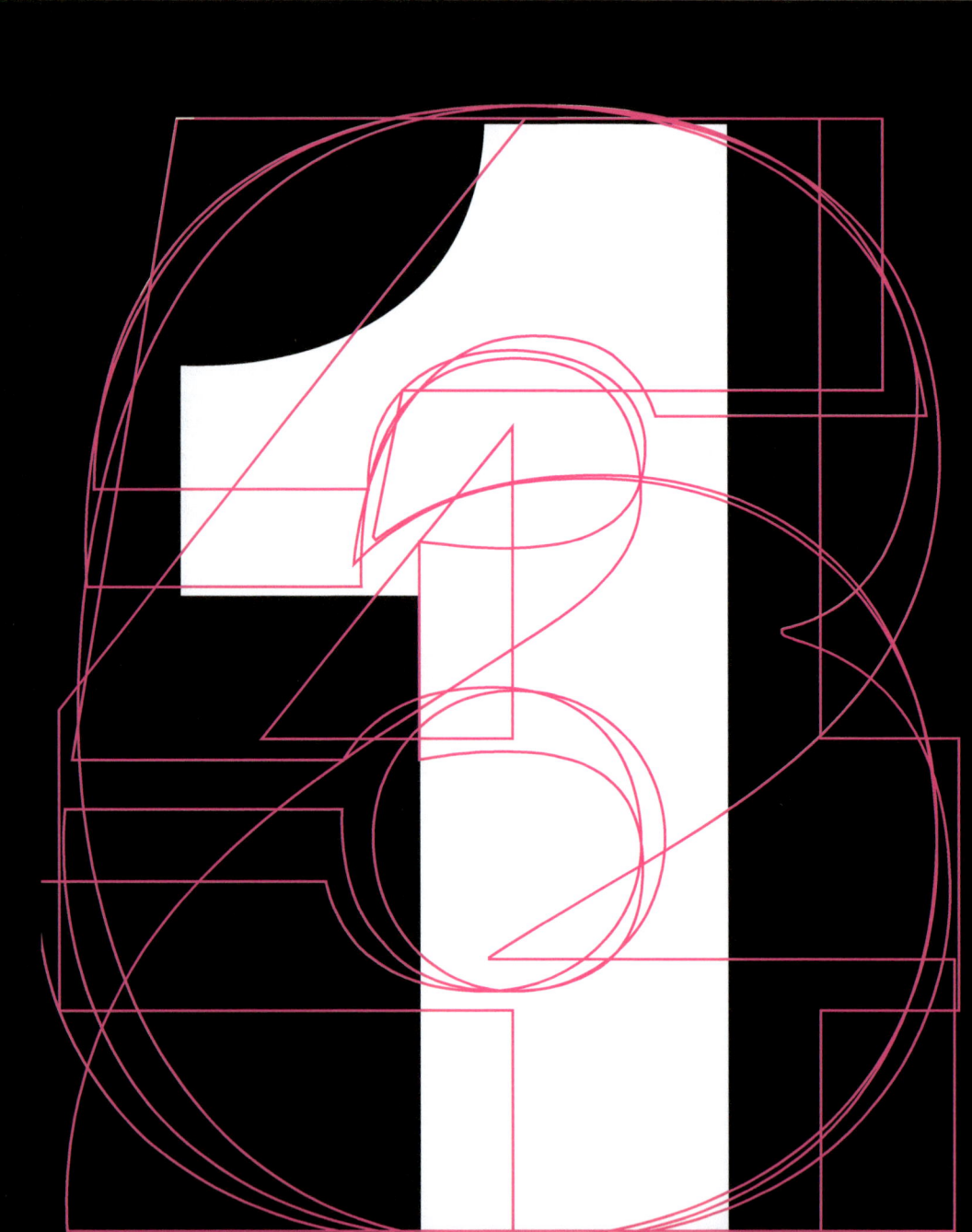

Just Imagine if it were Possible

Personal experiences contribute to a broad array of creative triggers that can influence a design journey. Continually absorbed, seemingly random encounters and experiences can become pivotal in identifying credible and compelling directions for the curious.
　　Embracing eclectic practices and appreciating that inspiration often assists with the generation of ideas.

> "You can't experience the experience until you experience it."
>
> **Bill Moggridge, co-founder of IDEO and Director of the Smithsonian's Cooper-Hewitt National Design Museum**

Experiences

Prior experience

Experiences that are encountered repeatedly create a knowledge base that can be called upon and used in the generation of ideas. The more experiences that an individual encounters, the greater and more diverse mental references become, all the better for suggesting lateral connections. However, the same knowledge can also become a hindrance if an individual is unable to redefine its meanings and is incapable of seeing objects in an alternative or imaginative context. It is always necessary to approach the design process with an **open mind** and to be prepared to experiment and appreciate objects in a new way. A closed mind only harbors a restricted, desolate, and unproductive imagination.

The imposing and compelling Mitate lamp collection created by Studio Wieki Somers echoes traditional Japanese culture, making individual references to subjects such as geisha girls and Japanese gardens. The multi-layered approach embraces these elements while also making the transition to create a contemporary collection in which the individual works have a specific lighting identity. The capacity to manipulate different perceptions and communicate an array of meanings within the complete collection is also evident at the component level, where familiar objects are used in unexpected arrangements and adopt a variety of contexts. The cross-pollination of culture and experiences captures the imagination in a manner that might not be so evident if the mix of references were absent.

Traditional craft, contemporary design thinking, and an array of experiences provided the platform for the inspirational collection.

Dutch designers Wieki Somers and Dylan van den Berg (Studio Wieki Somers) challenge convention in their creative approach to design. Collections including Mitate (2013) and Chinese Stools (2007) demonstrate an ability to engage in tangential thinking and to confront entrenched perceptions of everyday objects to reveal the previously unseen.

The Mitate lamps represent the principles of the samurai code of honor:

1. Gi (cord lamp) > right decision
2. Rei (shields lamp) > right action
3. Makoto (reflection lamp) > truth
4. Jin (fabric lamp) > compassion
5. Yuu (mirror lamp) > bravery
6. Meiyo (mesh lamp) > honor
7. Chuugi (black hole lamp) > devotion

Figure 1.1
Studio Wieki Somers, Mitate lamps, 2013.
Exhibited at Galerie Kreo, Paris.
Photo: Fabrice Gousset.

The juxtaposition of traditional and contemporary elements, the combination of purity and complexity, and the various ritual and convivial aspects of the lamps offer multiple opportunities for interpretation.

The ability to absorb and understand potential influences and subsequently rearrange and use them in an original direction is a trait that enables creativity, such as in Mitate, to emerge.

Figure 1.2
Studio Wieki Somers, Jin (fabric lamp). Photo: Fabrice Gousset.

Figure 1.3
Studio Wieki Somers, Mitate lamps, 2013. Detail of feathered pull-switch. The shrouded feather is revealed when illuminated. Photo: Fabrice Gousset.

Primary research

Understanding or identifying a problem usually requires an active "go do" approach to make important and meaningful connections and subsequently to place them in a relevant context. It is necessary not to be judgmental or biased toward any preconceived ideas, but to retain an open mind, absorb information, and be prepared to discover and accept the unexpected. Outcomes that contradict or challenge personal or broadly acknowledged views and understandings are of particular interest and should be carefully appraised.

Primary research can be conducted using a diverse range of methods, but the fundamental aspect should always be going out and asking questions, or observing direct from a reliable source (although this does not have to be limited to the most obvious source).

Assumptions can influence the process inappropriately, and so any initial approach should be broad and simple. It is the views of others that you are seeking, and not always the affirmation of a personal view.

Designing in isolation, without inspiration, and assuming that progress can be made without engaging in some form of primary research can be detrimental to the overall creative process.

The Chinese Stools collection could not have been created without direct observation of and interaction with, Chinese workers.

The **eclectic** street seating that is used, repaired, and crudely modified by workers in Beijing with access to very basic resources echoes the story of the workplace and the individual. The collection of stools with their curious appeal was purchased by Studio Wieki Somers and transformed into a more substantial aluminum collection.

The collection retained the individual identities and the ability to portray an intriguing story of street life.

Figure 1.4
Studio Wieki Somers, Chinese Stools – Made in China, Copied by the Dutch, 2007.
Photo: Pien Spijkers.

Journals and blogs

Creative journals and blogs provide a valuable source of reference and an array of opinions on contemporary issues. They direct and introduce ideas, engender contemporary values and thoughts, and reaffirm opinions on an international platform.

The articles and posts can stimulate unrelated ideas and directions if approached with an open mind. Journals and blogs are often visually focused and therefore encourage an international audience. Design is undoubtedly a sensory discipline, and in particular a visual discipline, and many triggers can be sourced from design journals. Even articles from overseas journals that an individual might not directly understand can still be inspirational and open up areas that have not previously been encountered. The potential to access journals and blogs authored by creative individuals with different cultural experiences and backgrounds is important, and it is not necessary for such publications to be directly related to a particular discipline. Probing articles in contrasting disciplines can readily trigger experimental thought and opportunities, since this becomes a forum of unfamiliar information and attitudes.

Accessing journals and blogs is not a replacement for primary research, but it is a necessary and accessible supplement in nurturing design ideas, and it can also become a useful time capsule for any subsequent retrospective analysis or comparison.

International, **interdisciplinary**, and **cross-disciplinary** journals and blogs provide a valuable source of visual reference, in which the discovery of unusual items and associations prompts tangential directions and proposes many lateral connections. Such exposure can introduce previously unconsidered thoughts.

For example, the elegant and fragile pink ceramic shoes documented by photographer and blogger Laura McCarthy suggest a range of conflicting characteristics, such as breakable plasticity, delicate strength, tough preciousness, and craftsmanship. The ballet shoes also prompt many sensory characteristics and visual directions that could ultimately inspire a very different genre of objects.

Influential design journals include:
Abitare, *Egg*, *MODO*, *ID*, *Metropolis magazine*, *Kult*, *FRUiTS*, *DEdiCate*, *frieze*, *dwell*, *icon*, *MONUMENT*, *INNOVATION*, *vanidad*, *domus*, *wallpaper*, *TWILL*, *mix*, *newdesign*, *Design Week*, *AZURE*, *surface*, *milk*, *FRAME*, *b0x*, *MARK*, *Design*, *intramuros*, *Blueprint*, *Artform*, *Aesthetica*, *Vogue Lowdown*, and *W magazine*.

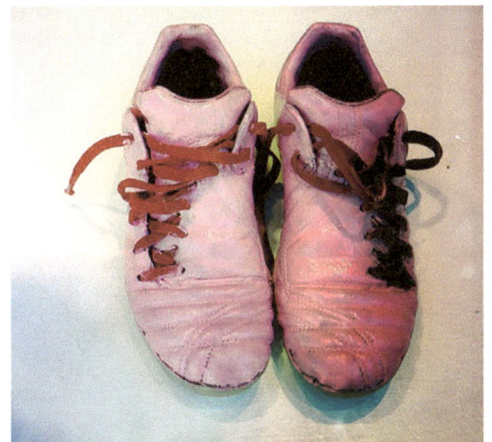

Figure 1.5
The Pink Ceramic Ballet Shoes.
Photo and blog: Laura McCarthy.

Scrapbooks, notebooks, and info dumps

Formulating a scrapbook is in many ways a personal and organic process. It should be a diary of creativity, a record of thought-provoking encounters that are created through the accumulation of interesting curiosities, an accessible place to store references.

Scrapbooks ultimately evolve into a comprehensive, tangible archive of thoughts where items of interest, which may or may not be directly related to a particular investigation, should be included as they have the potential to become a creative trigger and central to the development of an idea at a later stage. As with so many elements associated with idea searching, the development of a scrapbook and the collection of interesting **artifacts** need to be ongoing activities that capture and document experiences that fuel the imagination. Gathered references might be interesting for a particular characteristic and do not necessarily need to be considered in their entirety. It is often the accumulation of multiple references from very varied and seemingly unrelated sources that contribute to the eventual makeup of an original outcome.

The references in a scrapbook are often personal, and can individually tell a set of exclusive short stories. The gathering of a large number of seemingly eclectic artifacts from varied sources prior to, or during, the initial stages of a design journey creates the opportunity for an **info dump**—an arena of objects that can be openly discussed in the search for inspiration. Different objects, as diverse as flea-market ceramics, yard sale toys, or a kitsch plastic dancer may be informally discussed and considered within the info dump. The characteristics of the references are often more important than the overall reference.

Notebooks and sketchbooks further support the gathering of interesting observations, and need to be continually maintained to provide personal reference material, which is again capable of supporting possible idea generation.

The search for ideas and references that provoke creativity is a continual activity, an addictive journey for the inquiring mind, rather than a practice that only commences at a given juncture.

The process of initial idea generation is not always as simple as a switch-on and switch-off process. Capturing the idea prior to its departure is so important. After an idea has left, unless there is a visual trigger or prompt to assist its recall, it may be lost indefinitely. Since an idea is so potentially valuable it is imperative to secure it.

Figures 1.6a and 1.6b
Matthias Pliessnig, sketchbook entries.

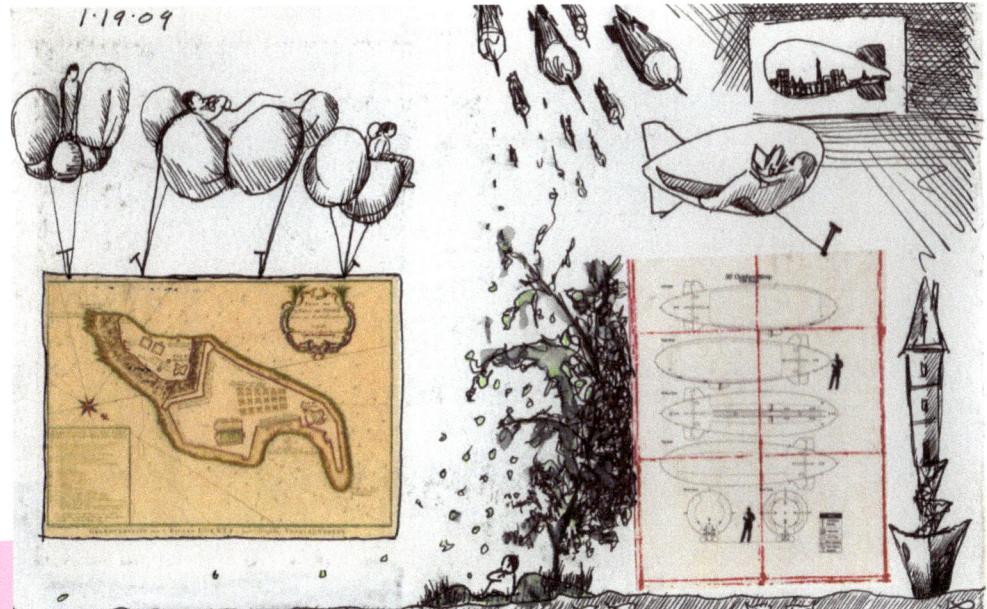

1.6a

1.6b

Notebooks

Simple, communicative marks created by artist-designer Matthias Pliessnig suggest a potential idea. The essence of the thought is beautifully captured with just a few lines and remains clean and pure rather than being stifled and overworked. Personal notes are scattered throughout the imagery and are balanced with more subconscious references and details. The creative shorthand guides the mind in a particular direction.

As multiple thoughts can arrive simultaneously they need to be recorded effectively, and so concise representations in a notebook provide the necessary trigger for exploring them later.

In addition to the ideas in his sketchbook, Matthias Pliessnig also explores initial thoughts through the creation of simple sculptures, which are referred to as "**adlibs**." Generated from basic objects found in the workshop such as pins, thread, crayons, and offcuts, the adlibs allow for thoughts to be explored physically and for the mind to play. The adlibs might not directly inform future developments, but as a creative activity they will undoubtedly make contributions to the practice of **lateral thinking**.

The combination of many triggers provides a valuable reference for development.

Figure 1.6c
Matthias Pliessnig, sketchbook entries.

Thoughts

Mental notes

Design is undoubtedly about thinking and understanding problems. The ability to look beyond the mundane, consider alternatives, and search for originality is fundamental to the overall process. An experience can be a particularly useful tool, but overexposure to something develops **mental baggage** or the inability to see things in any other context.

If something cannot be seen from a different standpoint it can become detrimental to the entire creative process.

When mental baggage is present it can be difficult to consider any other creative direction other than that of the overfamiliar. It is only when things can be thought of differently, without "creative blinkers," that it is possible to be original.

Operating beyond the normal confines of a discipline, gaining exposure to unfamiliar practices and accepting alternative values can assist in the ability to see or do something differently.

The opportunity to encounter diverse cultural experiences and to engage in dialog with other disciplines can steer the creative path around any preconceived thoughts, opening up doors of inspiration and possibility.

The thought processes are committed to solving problems through conscious and subconscious activity. Found, innocuous, and random artifacts, coupled with an element of play and a simple objective manage to temporarily redirect thinking to alternative challenges.

1.7

1.8

Figure 1.7
Matthias Pliessnig, Adlib: "write."

Figure 1.8
Matthias Pliessnig, Adlib: "strike."

"Imagination is more important than knowledge. Knowledge is limited. Imagination encircles the world."
Albert Einstein

Mental baggage

It is important to have a thorough understanding of a problem, rather than starting from a **hackneyed** or feigned standpoint. What if things have changed since an original idea was conceived, and successive generations have simply accepted the manner in which something is done? This question can be applied to so many aspects of the creative process, and to simply presuppose that what has gone before remains relevant, without questioning it, is difficult to accept in any circumstance.

Design aims to initiate rather than follow. The creative who is capable of questioning conventional attitudes is usually the creative who is able to set the benchmark for others to pursue, circumvent, or complement.

Watt?, designed by the renowned Paul Cocksedge Studio, echoes the studio's innovative design approach, questions conventional practice, and demonstrates the ability to simplify thinking. Watt? is different and yet beautifully uncomplicated, exploiting conductivity through the creation of a graphite circuit. A completed line switches the light on whereas a partially erased line switches it off.

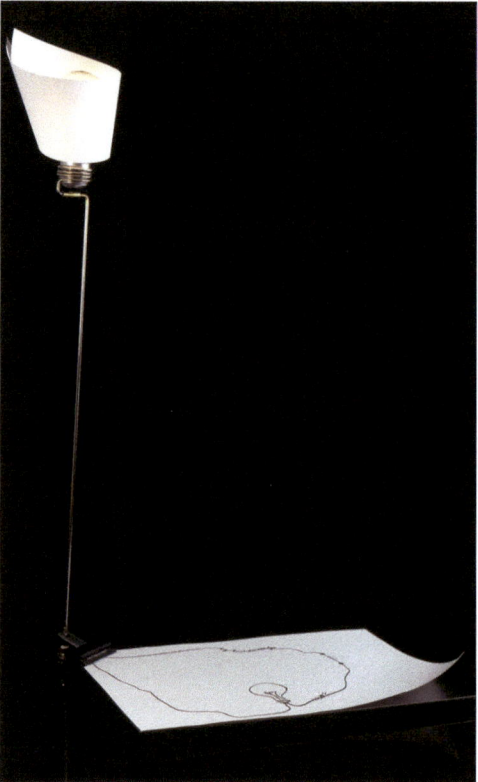

1.9a

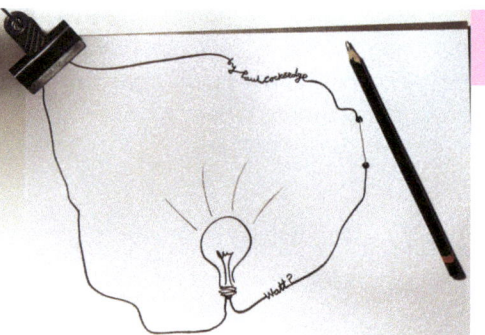

1.9b

Figures 1.9a and 1.9b
Paul Cocksedge Studio, Watt?
Photo: Richard Brine.

Mental baggage and the stagnation of original thought can occur for many reasons, but it is also something that can be overcome if our contexts and surroundings are adjusted, as the former director of industrial design at Philadelphia University and associate professor, Götz Unger, explains:

"I can get stuck for any number of reasons. I usually try to figure out what the obstacle is in my approach and then apply ways of getting out of a **creativity rut**. Here are strategies for my most common creativity ailments:

1. *Change the problem statement*
   A specific set of criteria for the envisioned design may lead to the same solution, especially if we have no new information to edit thoughts. Moving the problem definition around can lead to new opportunities.

2. *Get off the internet*
   Too much information and imagery is hard to manage; it can end up driving your work. Sometimes it is good therapy to rely on one's own intuition and to just "do it."

3. *Move around—to the coffee bar, to the park bench, or travel if you have the time*
   A change of scenery changes the way you feel about things. In situations where the distance between success and failure is very short, a small change in the way you think can be all it takes.

4. *Create order in your environment*
   A newly organized space is also a "new beginning" in the work you do. It is a mind game to get out of a rut and into an upward spiral.

5. *Persist*
   Put your thoughts aside and let time and a new context change the editing templates. Sometimes returning to a problem delivers the key thought that ties everything together elegantly.

6. *Ignore the problem*
   As far as detailing goes, sometimes the problems don't go away, they just get smaller.

7. *Visit an exhibit*
   Ideas transfer through cultures, disciplines, and times. It is a privilege to study the creative output of others, past and present. It is almost impossible not to take something away from an exhibit.

8. *Zoom out and then back in*
   When an existing design haunts you as the only solution, analyze how the design is conceived on an increasingly abstract level and then zoom back down on an adjacent path. Ideally, you will end up with a different object that has similar, desirable characteristics."

Figure 1.10
Ralph Ball and Maxine Naylor, Offset Vertical Cut, Anatomy Series, 2008.
Exhibited at "Indeterminate Cases," La Sala Vinçon, Barcelona.

Mental baggage and narratives are purposefully confronted with the Anatomy collection by influential designers Ralph Ball and Maxine Naylor. Familiar stacking chairs are dissected and deconstructed enabling alternative perceptions to emerge. What is the meaning of an object and how can the overall language be effectively manipulated and adjusted?

The alternative thinking Chair Anatomy collection formed part of Ralph Ball and Maxine Naylor's "Indeterminate Cases" exhibition (2008) at La Sala Vinçon in Barcelona.

Simple alterations to accepted principles and language can have a dramatic impact and create a beneficial springboard for discussion and further idea development.

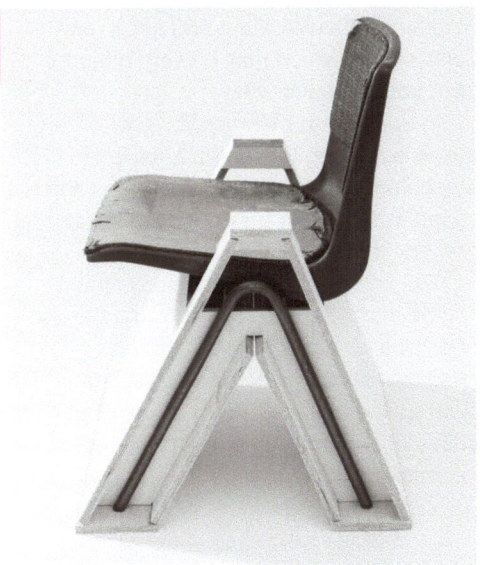

**Figures 1.11a, 1.11b, 1.11c and 1.11d
Ralph Ball and Maxine Naylor**, Archive Series, 2008. Top left: Full Case; top right: Back Case; bottom left: Leg Case; bottom right: Seat Case. Exhibited at "Indeterminate Cases," La Sala Vinçon, Barcelona.

Anecdotal references

Anecdotal stories from previous experiences and encounters provide an intimate and personal account of an experience and can be constructive in understanding the context of a situation.

A story that is told by many is capable of encompassing a wider range of issues. The observer has an important role to play in identifying the relevance of what is being communicated, and to grasp the core issues of the encounter.

A holistic comprehension can ultimately dictate, influence, or reaffirm an emerging idea. Simple wares can be a beneficial vehicle in the breaking down of social barriers and initiating a verbal or visual story, since they create a fundamental connection.

Design student Guan Ziyin (CAFA IFC 2014), innovatively instigated the telling of stories through the use of an old, wooden stool. Observing Beijing street life and stools comparable to those which Studio Wieki Somers used in 2007 to create the Chinese Stools collection, Guan Ziyin managed to rent a simple, wooden stool from a Chinese street worker, which then became the common link for a broad range of encounters and introductions.

Taking the stool to different areas, including the vegetable market and the park, she invited many different characters to sit on it and to have their individual stories captured in a photograph. The activity not only documented different individuals in their immediate surroundings, but also provided a substantial number of additional references, and a real, personal opportunity for informal and inspiring conversation.

Figure 1.12
Guan Ziyin, The Stool, 2014.

Simple practices such as those adopted by Guan Ziyin can have a significant impact on understanding real problems and accessing personal insight. As designers it is important to meet others, to talk with them and get a new perspective, which may be very different from your personal view.

In addition to being creative with the object being designed, a designer must also be creative with the many and varied approaches they adopt to obtain primary information.

The practice of capturing a moment in time and a simple story through the use of an object, especially if it also has character and a story to tell of its own, such as the street worker's stool, is an effective and enjoyable approach to information gathering.

Figure 1.13
Guan Ziyin, The Stool, 2014.

Figure 1.14
Guan Ziyin, The Stool, 2014.

Figure 1.15
Guan Ziyin, The Stool, 2014.

The formation of stories and the retelling of personal experiences are insightful in the generation of ideas. The Stool project by Guan Ziyin is in essence similar to the successful ongoing street art of international artist Luke Jerram and his beautifully simple concept "Play Me, I'm Yours," which has been circling the globe since 2008, connecting seemingly diverse individuals through curiosity and intrigue.

A simple idea and a simple approach or a simple idea and a global approach takes imagination and an intuitive understanding of what might be possible beyond conventional thinking. Artifacts that are used to bring people together, in the form of an info dump or in public spaces, may be as diverse as an old stool or a piano, but they can innocently facilitate many valuable opportunities for dialog that can subsequently evolve into innumerable connections, which might not have been possible otherwise. Anecdotal references are beneficial in generating inspiration but must also be appraised carefully as many different views are presented.

Luke Jerram constantly involves collaboration with artists, musicians, craft specialists, and the global community. The pianos have appeared all over the world since 2008 including Salt Lake City (2012), Hangzhou, China (2012), Paris (2013), and Luxembourg (2014).

1.16

Figure 1.16
Luke Jerram, Play Me I'm Yours, 2008.
São Paulo, Brazil.

> "If you are in a shipwreck and all the boats are gone, a piano top buoyant enough to keep you afloat may come along and make a fortuitous life preserver. This is not to say though that the best way to design a life preserver is in the form of a piano top."
> **Buckminster Fuller, *Operating Manual for Spaceship Earth***

Language

Observing alternative uses for everyday items can definitely provide a valuable reference for potential products, but as Buckminster Fuller pointed out, it is also important to understand a particular context and recognize when other pathways might be more viable or prosperous. An element of common sense needs to be continually present, but that does not mean that a particular characteristic of a surrogate item, an item that is used as a suitable substitute in a familiar **scenario**, is not beneficial to the process of idea generation.

An item that is given a responsibility in the absence of another specific item for a particular task is usually a good indicator of an opportunity. The following list identifies objects that have been observed being made use of due to their unintentional attributes, but which have responded appropriately to the imposed task:

Hat	as a	bowl	Friend's back	as a	desk
Hairclip	as a	bookmark	Tie	as a	belt
Rock	as a	hammer	Table	as a	shelter
Book	as a	fan	Cup	as a	pen holder
Paintbrush	as a	duster	Newspaper	as an	umbrella
Wall	as a	chair	Bag	as a	pillow
Table	as a	raft	Suitcase	as a	seat
Bed	as a	trampoline	Chair back	as a	coat hanger
Napkin	as a	sketchbook	Coin	as a	tin opener
Wallet	as an	album	Tray	as a	bat
Watering can	as a	shower	Bottle	as a	vase
Fallen tree	as a	bridge	Tree	as a	toilet
Card	as a	door stop	Lipstick	as a	pen
Boat	as a	bath	Violin case	as a	collection box
Scarf	as a	restrain	Garbage lid	as a	shield

The different and unusual ways that existing products are sometimes used can provide a useful insight into potential new directions. Insights are merely suggestions, rather than solutions, until designers such as Natalie Mao and Mars Hwasung Yoo take the initiative further and begin to consider, explore, and actively comprehend the process that needs to be adhered to in order to make the necessary transition from thought to sought.

The Chinese Hat (2003) designed by Mao Xiao-hua while at Tongji University, Shanghai, references conical Chinese hats and how they are sometimes used to contain simple items. The design is smooth, honest, and pure. It reflects the ability of the designer to control an idea, to demonstrate what Adolf Loos described as a "spiritual strength," and recognizes that outputs do not need to be complicated to be accepted and nor do they need to have unnecessary adornment to become viable. Controlling an idea and being able to manage a creative transition without being tempted to digress with unnecessary, resource intensive, detrimental additions is an essential attribute of a good designer and demonstrates confidence.

"Less is more," but the mantra is often overlooked and wonderfully pure ideas are all too easily lost through an inappropriate understanding of what is ultimately required and when a pure **aesthetic** is eventually tarnished.

Figure 1.17
Mao Xiao-hua, Chinese Hat: Ceramic Centrepiece. Exhibited at "Shanghai Made in Italy."
Photo: Natalie Mao.

The HAT lamp by Mars Hwasung Yoo or the Chinese Hat by Mao Xiao-hua might be considered simple, yet it is this simplicity of the works that makes the designs elegant, efficient, and appealing.

It is a mistake to drag out the design process and to continue to tamper with an idea unnecessarily. A designer must understand when to stop a design journey when the intended task has been achieved. So long as the designer has confidence in making a statement, the subsequent outcome will echo the original thought.

Confidence will come from understanding, acquired through continual observation, experimentation, and practice.

As Mars Hwasung Yoo states: "The form of the hat works as a shade but also connects us to many different images such as the beach, an elegant lady, specific scenes of a movie and more."

Figure 1.18
Mars Hwasung Yoo, HAT lamp, 2012. HAT effectively transposes the language of shade from hat to lamp. Exhibited at "SaloneSatellite," Milan. Photo: Stephanie Wiegner.

The terminology used to instigate design thinking can become an immediate obstacle if it is not manipulated effectively. Searching for abstract or alternative language supports the idea generation process in a similar way to how **analogous** objects can act as a **catalyst** for inspiration.

The ability to maneuver language, to shuffle contexts, and play with meanings is a simple but effective skill in the search for idea opportunities.

Considering the language that a particular cultural group uses or aspires to can assist in understanding the direction to follow. If the language currently being used is different to the language that needs to be used, the development of ideas can become sterile.

The potential for mixing ideas and creativity, from what might be considered to be opposite and unrelated backgrounds, is successfully demonstrated in the performance work of the international collective sampler-cultureclash. DJs, machine hackers, poets, and embroiderers, connect textiles and sounds where "sampler" is the common thread, and the foundation for experimental, creative activity that engages the imagination. Such collaborations might be unexpected, but when they occur they open the mind to an array of possibilities that are usually overlooked.

The embroiderer and the DJ might not be considered to have much in common but David Littler (sampler-cultureclash) demonstrates that the language they both engage in is remarkably similar. Discovering connections between seemingly disparate groups removes boundaries and creates the basis for a revealing dialog.

Embroiderer	DJ
Sampler	**Sampler**
Needle	**Needle**
Freestyle: as in embroidery	**Freestyle:** as in DJ/MC
Back: backstitch—a type of stitch where the thread comes toward you in a reverse direction	**Back:** back to back—DJ technique for playing the same part of a record on two different turntables to create a continuous loop
Cross: as in cross-stitch	**Cross:** as in cross-fader on a DJ mixer

The cross-stitch and the cross-fader both bring two different points together to meet and create a new union: one with threads and the other with sounds.

Mentally adjusting the given language for a design can effortlessly assist in the immediate generation of ideas, presenting an array of options from the outset. The Hubei Institute of Fine Arts (HIFA) adjusted the mentally suggestive term *chair* to *seating solution* during the StreetMaker project (2015), and in doing so promptly eradicated any preconceived images of conventional chairs that might have had a detrimental impact on the creative process. Such a simple change in the terminology openly invites a wider range of potential ideas to come forward. The ability to perch, lean, squat, and balance are among the many influences that can act as a catalyst for ideas associated with seating solutions. When such thoughts are combined with more traditional references, then additional ideas can also follow.

The design students at HIFA freely explored the locality to observe anything that could act as a possible seat, and sat on walls, posts, fences, stones, bags, buckets, oil drums, water coolers, market scales, bottles, rubbish bags, concrete poles, books, crates, tools, tyres, panels, rice bags, tubes, piles of bricks, timbers, wash basins, ceramic pots, rubble, playground rides, different aggregates, air conditioning units tethered to buildings, scaffolding, decaying branches, broken roof slates, and foam offcuts.

All the items identified offered a potential seating solution and an area of interest, although none of the objects would be immediately described as a chair. The students knew what to sit on through experience but might not have considered the obvious street references that continually surrounded them if the language adjustment had not changed their initial **mind-set**.

Figure 1.19
Designers at Hubei Institute of Fine Arts interview a broad range of individuals to identify possible seating solutions.

Adjusting the terminology of a problem can create inspirational triggers that may assist in idea generation.

For example, when designing a lunch box it might be beneficial to consider alternative terms for *lunch* and *box*. Although the objective remains the same, the introduction of a different language enables different thoughts to emerge.

Alternative terms for the word *lunch* (or associated concepts) and the word *box* (or types of containers) will conjure up a variety of options that can be coupled together to create potential new directions:

LUNCH	BOX
snack	pod
scoff	bag
brunch	sleeve
grub	envelope
eats	tray
fare	pot
nosh	bowl
nibble	cup
tidbit	bucket
goodies	basket
munch	case
mess	plate
feast	pack
morsel	carton
pickings	bin
bite	crate
gorge	chest
wolf	packet

The imaginative terms for *lunch box* might therefore become:

> Gorge bag
> Feast tray
> Mess bucket
> Scoff box
> Nibble bin
> Pickings pod
> Morsel case

The introduced terms manage to steer thinking away from any preconceived notions or expectations and allow the idea process to become more creative.

At the outset of the design thinking stage the mental adjustment of language can be a beneficial tool.

Analogous

An analogous object is an object than can assist in the generation of an idea, and it can come from literally anywhere. It is an analogy, or something similar or comparable, but not the same.

Although the analogous reference might emerge from a seemingly unrelated genre of objects, if it shares a comparable characteristic, no matter how different or random the object, it can become an essential influence in mentoring an emerging thought. When multiple analogous references are identified, from multiple arenas, any new proposal can become dramatic and exciting.

If references are only historical, from within the same "gene pool," then there is a greater likelihood that the conclusion will be more **banal** than original. Analogous references are important in introducing alternative solutions.

Figure 1.20
LaLa Lab, Fire, 2011. Exhibited at "SaloneSatellite," Milan.
Photo: Courtesy of Yuki IIDA, LaLa Lab.
The Fire lamp presented at the "SaloneSatellite" (2011) by Japanese design laboratory LaLa Lab is both analogous and inspired by an open fire. The connection manages to encompass many aspects of the original reference.

The search for an analogous reference is simple if things can be taken out of context and seen differently. The ongoing cross-pollination of references through exposure to diverse experiences supports the identification of analogous objects.

Ronan and Erwan Bouroullec, considered to be amongst the most influential of designers in European design since the presentation of their Disintegrated Kitchen project at the 1997 Salon du Meuble in Paris, have created designs for many of the leading companies in global design including Cappellini, Kartell, Established & Sons, FLOS, and Magis.

An inherent ability to identify creative connections in multiple disciplines ensures that the design outcomes are unmistakably distinct and captivating.

Analogous references use a familiar language but are often different enough to delight an audience when presented in an alternative format.

Figure 1.21
Ronan and Erwan Bouroullec, Quilt chair, 2009. Created for Established & Sons.

> "If you have an apple and I have an apple and we exchange these apples then you and I will still each have an apple. But if you have an idea and I have an idea and we exchange these ideas, then each of us will have two ideas."
> **George Bernard Shaw**

Brainstorming

An idea can arrive, unannounced, at any time, but it can also be encouraged. It is often assumed that the origin of an idea is initiated with a **brainstorming** session: a process that adopts a variety of formats but is essentially where a trigger stimulates the announcement of a potential idea within a creative arena. But such a session relies on individuals having already encountered a diverse range of experiences that are related to the prompt in a direct or indirect manner.

As previously mentioned, the development of an idea, although not necessarily realized, usually begins much earlier and may remain dormant until a particular creative activity, such as a brainstorming session, introduces a suitable verbal or visual catalyst. The brainstorming session might be considered therefore to be a vehicle for releasing experiences, stimulating the imagination and making connections.

A brainstorming session that engages a few individuals with divergent backgrounds, creative experience, and open minds has the potential to be more beneficial than an open forum where ideas are too easily rejected.

There is a certain irony that brainstorming sessions often adopt a hackneyed process, using sticky paper squares to capture ideas. The initiation of the brainstorming session often seems to forget what creativity is, and that the overall process is actually a creative activity in itself. It is the gathering of select imaginations that is more important, rather than the sticky square, and many leading design offices have very successfully circumnavigated the dilemma.

Hy Zelkowitz, a sculptor-designer based at Philadelphia University, USA, sums up the problems surrounding brainstorming in a succinct manner:

"Basic misconceptions about brainstorming: one is that an idea can fit on a sticky yellow square of paper. Another is that the process is to show quickly why many ideas won't work. Then, of course, there is the preposterous notion that after I have just had an idea, what I really want is to be distracted by discussing other people's ideas. And lastly, we seem to think the more brains the better.

"Smaller brainstorming groups, four or five people, can address all of these by providing more time for a thorough, generous, and patient exploration of everyone's ideas and the opportunity to play with and build upon those ideas."

> "It seems a very dangerous idea. It is – all great ideas are dangerous."
> **Oscar Wilde**

Brainstorming sessions condemned to a space devoid of inspiration are to be avoided. Design studios are often a sanctuary for unconventional peculiarities sourced from varied encounters, and provide a beneficial and creative backdrop for stimulating the imagination.

A brainstorming session can be orchestrated to encourage the imagination to search for unforeseen and unexpected depths.

An innovative approach that global design group, Lunar, adopts to stimulate creative thinking, is to show that connections can be forged between very abstract subjects.

To demonstrate the process, random themes, such as a cat and a fridge (Future Cool, 2012), were selected to show how many similarities could emerge from such disparate subject matter. Although contexts change and subject matters can be more tailored, even a cat and a fridge can be seen to have connections in the sense that they both contain milk, both scratch the floor, and both purr. Connections are there to be discovered.

A capacity to initiate such a discussion and to search for unexpected associations with unexpected subjects opens the mind and provides an ideal foundation for any subsequent idea generation activity.

The sharing of an idea is frequently beneficial in a bid to generate different scenarios and directions. An idea that may appear absurd might become the motivating factor for another's input into the process. Ideas should be encouraged and allowed to develop rather than being dismissed too rapidly due to insufficient consideration.

> "Brainstorming in a group is a great way to change the direction of one's thinking. Manage a session for ten intense, fast-paced minutes and then mine the comments and ideas for value. Combine ideas, follow the trajectory of some and see where they take you, let the half-baked thoughts seed viable solutions."
>
> **Götz Unger, Director of Industrial Design, Philadelphia University**

Designers at IDEO tackle projects through sharing experiences and embracing an inherent aim to understand the essence of a given problem through assorted and tailored practice. Creative approaches are able to revoke mundane and unimaginative habits associated with initial idea generation in preference to the creation of ingenious and practical scenarios. A tangible outcome with meaning and context, where connectivity is realized rather than isolated, is constructive in the brainstorming process.

Many of the world's leading design studios continually adjust and adapt the details of their brainstorming activities depending on the particular context. Although this is to be expected, for others a brainstorming session simply appears to be an automatic stage, a mental process that flags up the predictable. Real inquiry requires real thought and careful consideration on how to conduct different sessions. The objective should always be to illicit ideas in a manageable and explorative manner and to provide sufficient opportunity for ideas to breathe. It is understood that different problems require different approaches. A space with character and an array of visual references, not necessarily related to a particular theme, will feed the senses.

1.22

Figure 1.22
A multidisciplinary IDEO team prototypes the flow and layout of a space using lo-fi materials.
Photo: IDEO.

The facilitator of a brainstorming session needs to retain an open mind and be aware that pursuing a particular line of inquiry before disregarding it is often a positive trait.

The approach to brainstorming should be carefully coordinated and considered, building on experience and an understanding of how to extract ideas effectively. The actual session can be an opportunity for the imagination to flourish. A brainstorming session is not limited to the outset of the design process, but is an activity that can be continually revisited and modified, as an idea is nurtured.

Design studios such as IDEO, Lunar, and Astro Studios are havens of creativity, where ideas are continually considered and given the respect they deserve. What might appear as an eccentric or outrageous concept can readily evolve into an unconventional or unusual proposal with a very plausible future in the right hands.

Figure 1.23
Photo: Courtesy of Astro Studios.

Observations

Improvisation

Recognizing the solutions that individuals create in everyday life to solve immediate and often personal dilemmas provides a useful reference for the generation of ideas.

Comparable problems are solved in multiple ways depending on the resources available and the creativity or ingenuity of the individuals tackling them.

Random items are often used in a manner for which they were not originally intended, but these diverse applications often create intrigue and an opening of the imagination.

Improvisation allows for rules to be broken and for metaphorical shackles to be discarded in preference over naivety and creative innocence.

Improvisation is personal freedom, and should be enjoyed when it presents itself as a possible indicator for development.

Inner-city street seating often provides a valuable insight to creativity where impoverished materials solve immediate problems. Stools covered in rags held tentatively in place with scraps of rope, broken chairs with legs crudely wired together, and plastic seats precariously balanced on broken crates are common. Such works are brutal and creative, harsh but innocent, and loved but ignored. They undoubtedly provide valuable inspiration to those who "see" with the eyes of Henry David Thoreau.

1.24

Figure 1.24
Street chair, Nanchang Street, Wuxi, China, 2015. Bleak and often severe repairs increase the life of a street chair but can also silently suggest thoughts for development.

"IDEO think of **prototyping** in three main phases: inspire, evolve, validate. The inspirational phase is the right place to try out ideas by making things, to use low-resolution techniques, and to embrace failure."
Bill Moggridge, co-founder of IDEO and Director of the Smithsonian's Cooper-Hewitt National Design Museum

Producing a "dirty" or "junk" model is an effective process for evaluating initial thoughts, and significantly assists in the immediate communication of a vague idea to others. Readily available, non-specific objects become improvised materials, which are integrated into the emerging form.

The items do not need to be special, but rather an approximation of what is needed, and are assembled using available means. The constructed ideas can be interchangeable to effectively explore alternative directions. There are no particular guidelines in the generation of such a model except that imagination and creativity should be used to the full. A dirty model is not created to be particularly beautiful or elegant, but rather it is created to be informative, provocative, and efficient.

1.25a

1.25b

Figures 1.25a and 1.25b
A creative workshop at Tsinghua University, Beijing, China (2012), explores what IDEO describe as an inspirational phase by exploring a variety of design ideas through **junk modeling**. The process allows for rough representations to effectively communicate an initial idea.

Observations

> "Design is for me observing the world, analyzing meanings of objects and giving visual comments."
> **Gijs Bakker, Gijs Bakker Design**

Dutch designer Tejo Remy created the three-part series Milkbottle Lamp (1991), Ragchair (1991), and Chest of Drawers (1991), which contributed to the first Droog collection. The designs use found objects and tell their own unique story. The works can be seen to have an affinity with solutions that were initially founded on need.

1.26

Figure 1.26
Tejo Remy, Droog, Rag Chair, 1991.
Photo: Gerard van Hees.

> "The pursuit of the homemade paradise. Like Robinson Crusoe stranded on an island he had to reinvent his new home with what he found and what was to hand. The Milkbottle Lamp, Ragchair and Chest of Drawers are conceived from this idea. Making things with what is available, **up-cycling** material to create items with a better life or meaning, in a world where resources are ending."
> **Tejo Remy**

Tejo Remy's Chest of Drawers (1991) is one of the most iconic pieces of design from the 1990s. Created from salvaged drawers, each with its own individual story to tell and place in history, the work demonstrated how seemingly eclectic components—when configured as a collective—could become a work that is able to challenge contemporary thinking. The work, manufactured by Droog, formed part of the initial Droog collection in 1993.

Figure 1.27
Tejo Remy, Droog, Chest of Drawers, 1991.
Photo: Bob Goedewagen.

1.27

Observations

> "The project (100 chairs in 100 days) suggests a new way to stimulate design thinking, and provokes debate about a number of issues, including value, different types of functionality, and what is an important style for certain types of chairs."
> **Martino Gamper**

The 100 Chairs in 100 Days project by sculptor-designer Martino Gamper used abandoned furniture and objects to create a collection of chairs, each with an individual identity. The collection includes a broad range of objects and solutions, such as the crossover of materials in Mono Suede (2005), the diversity of materials in A Basketful (2006), and the combining of hard and soft elements in Barbapapa (2006). The need to make decisions, blend components, and stimulate potential ensured that the collection challenged the dialog of design.

This application of generally detached components to generate alternative narratives is also examined in the Off-Cut collection (Martino Gamper, 2011).

Good design thinking does not require good materials, and with imagination many rejected materials can be beautifully and successfully used.

The ability of something to change from the ordinary to the extraordinary is an exciting direction to explore. The metamorphosis of an object may be an incidental outcome but might also be a permutation initiated by constraint.

"I have always been fascinated by the raw musical power that an orchestra can express, and after creating a series of videos where I performed a multi-track piece with a single custom-built instrument, I decided to take the concept a step farther and create my own orchestra made of unusually unique instruments."
Diego Stocco

Italian composer, musician, designer, and sound innovator Diego Stocco discovers and manipulates sound through the creation of innovative instruments that, in a similar approach to Martino Gamper, combine basic hardware, redundant or broken objects, and simple found items.

1.28

Figure 1.28
Diego Stocco, Custom Built Orchestra: Experibass, 2012.
Photo: Gianfilippo de Rossi.

As the creator of the unique Custom Built Orchestra, the instruments that Diego Stocco constructs are as original as the sounds they produce.

The collection originally developed with the Experibass, a **hybrid** instrument that used the body of a double bass to support characteristics of the string family including elements of a violin, cello, and viola.

The manipulation of the structures and subsequently the sounds of these instruments manages to challenge conventions, awaken the senses, and stimulate further thinking among audiences.

In addition to the Experibass, the Custom Built Orchestra includes the Experiviolin, a structural and musical amalgamation of a violin and an electric guitar incorporating acoustic strings, the culturally diverse Expericello, where the base of a broken cello is salvaged and combined with a zither, and the electro-acoustic Arcophonico, which merges a found branch with an array of string-based instruments.

The interaction of the instruments is also often an original activity, and in the case of the Arcophonico it can be played using an assembly of practices and items ranging from pinching to the more abstract adoption of a chopstick.

Experimenting with sound, Diego Stocco embraces a methodology that explores many unconventional approaches through the ability to question and ask "what if?" Luminopiano, an instrument that captures and magnifies the sounds of the tungsten filament, exemplifies an improvisational approach, as does the percussive instrument Tonal Metals, which uses basic kitchen hardware. Ideas behind proposals such as the Custom Built Orchestra have been able to wander and develop, and in doing so test familiar opinions related to the understanding of how sounds and music are created.

As with most objects, damaged and abandoned instruments have so much potential when viewed differently or away from their original context. A change of mind-set from what is often too readily accepted can rapidly offer up many previously unforeseen opportunities.

The use of analogous references, as previously mentioned, coupled with an ability to improvise, creates the innovative foundation.

Figure 1.29
Diego Stocco, Custom Built Orchestra: Textural Flute, 2012.
Photo: Gianfilippo de Rossi.

1.29

The Textural Flute exploits the restrained and confined movement of air. The consolidation of basic tubing, penny whistle, and a trombone in an arrangement that shares an analogous language with many instruments creates an original instrument and a unique sound.

Observations

Figures 1.30a and 1.30b
Mark Zirpel, Water Organ, 2011.
Photos: Mark Zirpel.

Constraints

Why do something ordinary when an astonishing and memorable challenge can be set? Audiences frequently seek originality, entertainment, and innovative experiences rather than more of the too-familiar.

The challenges of Martino Gamper, Diego Stocco, and others are also demonstrated in the multidisciplinary installations and outputs of Mark Zirpel.

Mark Zirpel's ability to ask poignant questions, questions that search, drill down, and have meaning are key to the glass sound sculptures Water Organ and Rain Organ that manage to successfully blend art, science, and performance through displacement to create a wonderfully unique sound.

The automated ascent and descent of a collective assembly of **bespoke** labware-like vessels instigates the movement of the water, causing air to be forced through various outlets, analogous to more familiar orchestral instruments, to create many exclusive sounds.

Posing taxing questions and setting challenging and demanding constraints ensures that something original will be revealed.

The works—a combination of labware, tubing, stoppers, wires, glass horns, and whistles—do not conform to what might be perceived as an instrument, but nor should they. These original works open the mind, allow others to be introduced to alternative thinking, and can become a stimulant for further thinking where different constraints might be introduced. Beauty and practicality naturally support each other.

> "What's the monobloc chair worth? Can you add substance to an industrial product by taking away material, bit by bit, hole by hole, eventually rendering its structure too fragile to support its traditional usage? Will we perceive the monobloc, once we can see through it?"
> **Tina Roeder**

Transformation

The White Billion Chairs project by designer Tina Roeder (2002/2009) is a limited and numbered edition of thirty-three unique pieces. The chairs are individually perforated with up to 10,000 holes and sanded by hand.

The design challenges the juxtaposition of adding through removal. Such questions are necessary to explore meaning and to appreciate that there is often more to an object than the object itself.

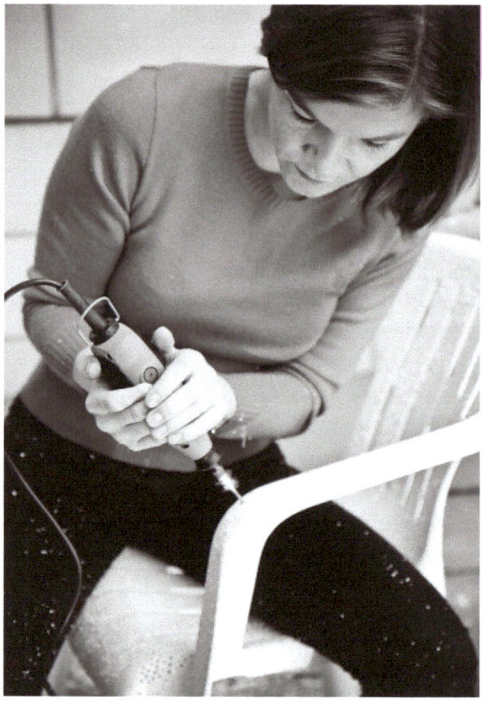

Figure 1.31
Tina Roeder, White Billion Chairs 2002/2009.
Exhibited at Appel Design Gallery, 2009.
Photo: Guido Mieth.

> "Texture of erratic nature stands in strong contradiction to the shape. The object embodies the relationship of industrial vs. natural."
> **Studio Libertiny**

Artist Cat Chow created Studio Recordings (2006), a series of large-scale coils carefully assembled from various reclaimed belts. The ability to see the potential of unfamiliar variables and orchestrate with meticulous execution is a trademark of Cat Chow and provides a valuable **benchmark** when questioning the possibility of objects. The work of Cat Chow radiates quality and detail. In addition to Studio Recordings, Cat Chow has created works from Power Ranger trading cards (Power Ranger Kimono, 1998), to zippers (Undress, 2004) and reclaimed keys (Keeper, 2008). The simple, readily encountered items are removed from their familiar contexts into a completely different field through the addition of time, sensitive craftsmanship, and understanding.

Studio Libertiny, renowned for experimental and creative approaches that realign conventional thinking, such as The Paper Vase (2007) and The Unbearable Lightness (2010), explored the possibilities of welding in 2009. Coupling the diverse practices of craft and industry with the addition of a robotic system enabled the studio to create the profound Weldgown from coiled layers of welded stainless steel.

As demonstrated in the works of Cat Chow, it is an attention to the intrinsic details that is so fundamental in the creation of Weldgown.

Figure 1.32
Tomáš Libertiny, Weldgown, 2009.
Photo: René van der Hulst.

Observations

Figure 1.33
Ralph Ball and Maxine Naylor, Plastic Gold, Archaeology Series, 2004.

A manipulation of perception is evidenced in the work Plastic Gold by Ralph Ball and Maxine Naylor. The omnipresent monobloc chair, an object that initially broke creative boundaries but that has subsequently become so familiar that it is no longer distinguishable from its surroundings, is transformed from being insignificant to being meaningful.

The approach is also echoed in the work of Dominic Wilcox (Luxury Skimming Stones, 2009), where altering a perceived status immediately poses a dilemma as to if and when the stones should be used and lost to the depths of the water.

The ability to see the unseen and to consider a familiar object from an alternative perspective supports creative thinking.

Figures 1.34a and 1.34b
Dominic Wilcox, Luxury Skimming Stones, 2009.
Photo: Dominic Wilcox.

> "I get inspiration about my design research and projects by people, by their use and habits, by their way to be collaborative and innovative and by the design attitude they have even without being designers."
> **Davide Fassi, Associate Professor, College of Design and Innovation, Tongji University, China, and Assistant Professor, Dipartimento di Design, Politecnico di Milano, Italy**

Inspiration

Inspiration is everywhere, especially if objects are considered out of their immediate context. An open mind allows for things to be thought of differently. The removal of an accepted "norm" that circulates around a familiar object immediately suggests that something original is going to be announced. Deliberately applying constraints to objects to explore alternative options can reveal extreme scenarios, and in doing so can prompt an unexpected line of inquiry. The method for connecting abstract ideas used by Lunar typifies such an approach, and is widely used to discover unique thoughts.

Considering the opposite of what something is "expected" to be, such as the Transparent Chair (2011) by Oki Sato at Nendo, plants a seed of thought and sets a creative standard.

Searching for ideas is not always a journey to find a solution to an existing problem, but rather sometimes a journey to find a different problem, a problem that hasn't previously been considered.

Adjusting the circumstances that surround an object and exploring the unusual and the curious is worthwhile and can lead to developments that turn social attitudes.

A questioning mentality regarding an existing object, process, or material to deliberately find an alternative use will unveil plenty of directions, even if at the outset it is not clear what it is that the designer is looking for. Artifacts and resources do not need to be categorized according to their history, but rather their potential, and this can only be understood if the right questions are continually asked by a curious mind. An opportune discovery is not necessarily a solution but perhaps just a single option to be considered.

The search for ideas is a constant series of questions.

"A chair made with polyurethane film, a transparent film commonly used as a packaging material for precision instruments and products susceptible to vibrations and shock, thanks to its high elasticity and ability to return to its original state. Looking at the chair, it seems to consist of nothing but a backrest and armrests. It wraps and supports the body like a hammock, providing a light, floating feeling for the sitter."
Nendo

1.35

Figure 1.35
Oki Sato, Nendo, Transparent Chair, 2011. Created for Milan Design Week.
Photo: Masayuki Hayashi.

Observations

Although many products are mass-produced, it is often the case that an item has been developed in such a way as to give the impression that it is unique to the user. Despite achieving the feeling of belonging, many users will still make further adjustments to personalize a design. The reasons for alterations may at times be unclear, but for whatever reason the user felt them necessary. It is important to observe such alterations and to consider why they have been made. Exploring and understanding these user refinements provides valuable inspiration.

Inspiration is everywhere and the search for ideas and inspiration is a continuous activity. An idea can arrive in the most unlikely situation, and yet when it does it cannot be ignored. It is necessary to be continually looking and seeing the activities and interactions in our immediate surroundings and trying to appreciate the various meanings and reasons for what is being observed. Engaging the imagination to immediately offer alternative contexts for such everyday observations, in an attempt to prompt potential opportunities, is of fundamental importance.

An observation might not be immediately relevant to a particular task but it needs to be recorded for reference.

Kelly Chen observes a worker in China taking a rest and using his hat as a makeshift mat to sit on in damp conditions.

Following a short break the worker gets up and moves on, placing the hat back on his head. The activity can't be ignored, and although the worker's hat might not be an ideal solution to his particular need, the observation does indicate that there is a problem, which might forge a series of scenarios to prompt further thinking.

Figure 1.36
Using a hat as a makeshift mat.
Photo: Kelly Chen (2015).

Performance

Ceremony can be defined as a distinct procedure, a performance that adheres to protocols and rules. A ceremony can often adopt a theatrical characteristic and is reserved for special occasions or for creating a sense of importance. Although usually associated with religious experiences, rituals are encountered all the time and are often conducted or acted out subconsciously by an individual or group.

Altar of Things, designed by Stephan Schulz for the Calvin Klein Home Collection, recognizes the importance of ceremony and formal performance. The elevated and controlled aesthetic of the design ceremonially offers up artifacts. Beautifully simple and controlled, eclectic objects are afforded an additional importance when organized respectfully on the altar.

A recurring theme, but the ability to control design demonstrates a confidence that reinforces the ceremony.

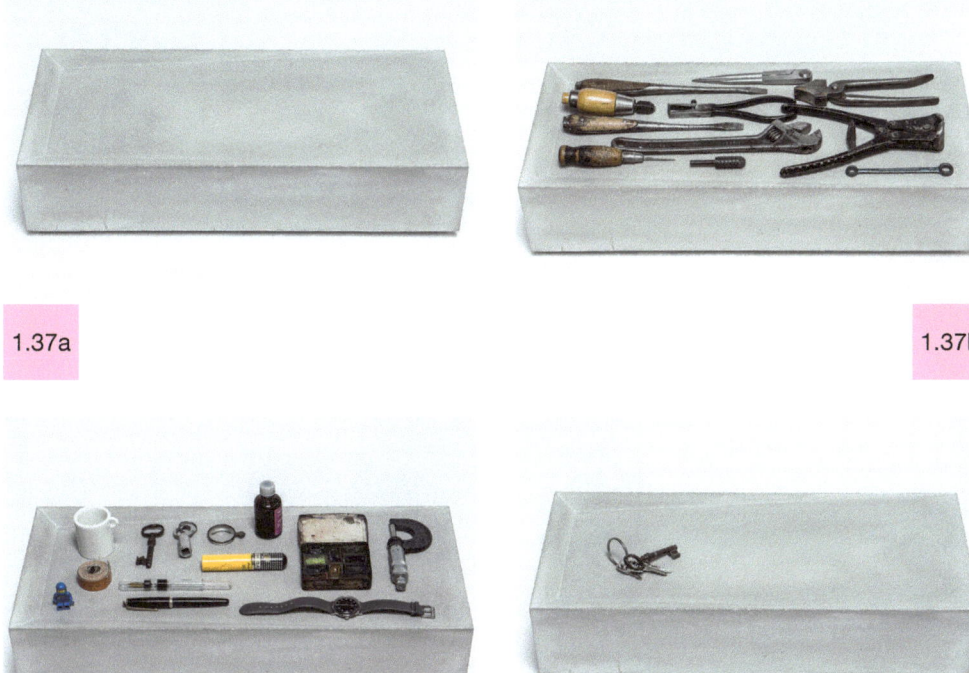

1.37a

1.37b

1.37c

1.37d

Figures 1.37a, 1.37b, 1.37c, and 1.37d
Stephan Schulz, Altar of Things, 2013. Created for Calvin Klein Home Collection.
Photos: Matthias Ritzmann.

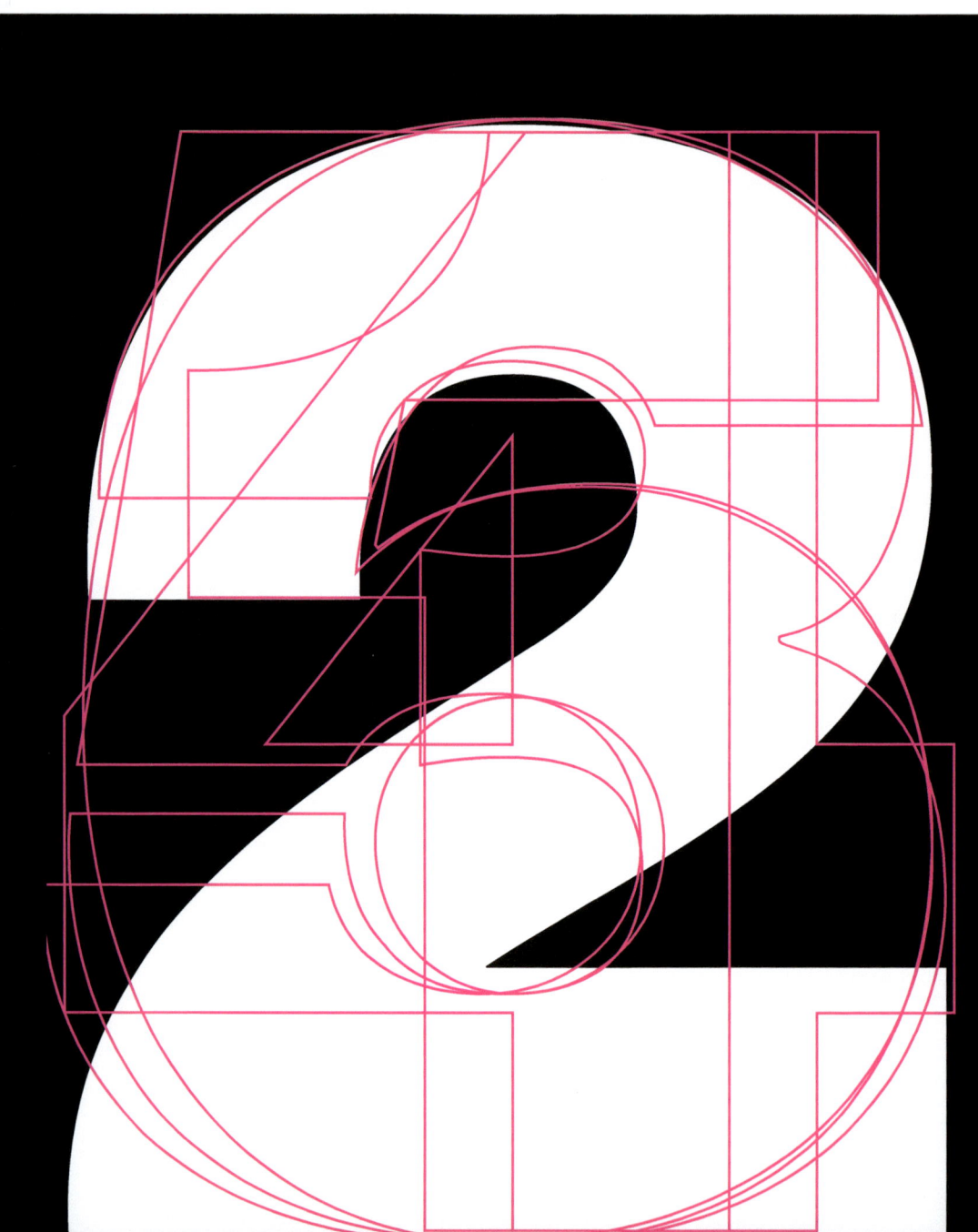

Thinking Differently

Creativity needs to be allowed to explore and wander with abstract narratives without bias but conversely must also operate from an appropriately informed foundation. Self-imposed constraints, profiles, themes, and historical references all influence and guide the creative process. Challenging conventional attitudes is a continuous and necessary activity that can reveal many varied opportunities.

Role play

Gaining an understanding of something can be achieved in many ways, and often involves some form of interaction or dialog with others.

A simple way to get an immediate impression of an idea is to use **role play**—the raw acting out of an activity or problem—as this can provide an effective opportunity to identify fundamental issues that might otherwise be overlooked. If such issues are not identified from the outset, then it is possible for a project to be initiated on an inappropriate path.

This activity is essentially a chance to relate directly to and to interact with an emerging idea, or to appraise an existing scenario. It is akin to a childlike performance that is heavily reliant on the imagination.

The practice provides an alternative and simple way to see things, and can suggest possible directions to follow, which might not be evident if the problem were approached in isolation.

Mental role play can be conducted at any time and anywhere by an individual manipulating a particular thought in their mind, and can be supported with small movements or perhaps ambiguous sounds. The movements or sounds can become necessary prompts for the thoughts being considered by the individual. This activity might appear strange, but it has an important part to play in idea development.

Physical role play—the acting out of a situation—can be approached in a space where nondescript items, including colleagues, can become essential props. The activity might be an individual or group process, but it provides a framework for the imagination to relate to. The performance may appear vague to disconnected observers, as it is only the **literal thinking** connections that can be viewed, but when combined with active thinking many questions can be answered and directions can be encouraged or dismissed efficiently. The props might be as basic as a colleague occupying a particular space to envisage how a particular object of a similar volume might appear, or they might be a few found items that echo a particular form in a very loose manner, but enough to feed the imagination.

The imagination makes a connection in much the same way as it does when a child plays. The process of role play is often overlooked, but it remains a very effective process in identifying some of the core issues.

2.1

Figure 2.1
Adam Verity, 2008–2014.

Purpose

The purpose of an object can appear to be obvious but it is not always that simple. A street lamp lights the area around it at night, but as with most things it is also used in a great many other ways.

Observing how an object is used and how a community actually interacts with it can reveal many more functions and perhaps some suggestions for development.

The street lamp might light the street but it can also be used:

- to hold up a particular sign
- to suspend or attach a traffic light to
- as an emergency barrier for something to crash into
- to lock a bike to or lean it against
- to tether a dog to
- by spectators at an event who climb it to get a better view
- to display a poster for a local event
- to connect flags and decorative items to
- for birds to perch on
- to stop and rest against
- for friends or family to identify as a meeting point
- for children to use in a street game
- to lean on
- for the graffiti artist to tag
- to lean against when sitting on the ground
- to tie a temporary barrier to
- to attach a bin to
- for electricians to connect cables to and from
- as an architectural feature.

Each of these uses can be considered in the design and can be either accommodated or designed out.

Preventing a dog or a bike from being attached might simply require creating a base that is too wide to tie something around. The same approach might prevent individuals from climbing it. A ridged base might prevent someone from leaning on it, or a textured surface may prevent items being attached to it. The design does not need to be a vertical pole: it might be contoured near the base to encourage sitting, or it might be a vertical light pole rather than a pole with a light at the end. The light might be used to project a particular message or sign.

There are many possibilities for what the street lamp could become and what it doesn't need to be.

Assumptions

The imagination has the capacity to see things that aren't there; it fills in the gaps to create an impression.

The Eigruob table lamp designed by Oki Sato at Nendo (2014) is a demonstration of lateral thinking and a move away from what might be considered an obvious or assumed response. Nendo looked beyond the obvious and at the space created between two lights when designing Eigruob for Italian manufacturer Kartell. Eigruob was designed to celebrate the tenth anniversary of the original Bourgie table lamp designed by Ferruccio Laviani (2004).

Figures 2.2a and 2.2b
Oki Sato, Nendo, Eigruob, 2014. Created for Kartell to celebrate the tenth anniversary of the Bourgie table lamp (2004) by Ferruccio Laviani.
Photo: Akihiro Yoshida.

The silhouettes and the transparent characteristics of the original Bourgie lamp provided the inspiration for Eigruob. Nendo designed a lamp inspired by the negative space between two Bourgie lamps, which is identical to the inverted version of the original lamp, and in doing so paid homage and respects to the 2004 Laviani design without actually reconfiguring it. The name Eigruob is the reverse of Bourgie.

Making assumptions is dangerous, and should be avoided or at least approached with caution. A person's appearance or the perception of an object can frequently be far from accurate on closer inspection.

A third party who is not experienced often makes judgments that are incorrect due to misunderstanding the particular context or situation.

Assumptions are usually wrong to some extent, and unfortunately, if given the opportunity to develop, can instigate a series of inaccurate thoughts and associations.

References to **stereotypes** are damaging and difficult to ignore, in a similar way to how mental baggage relating to an object can hinder a pure judgment.

Figure 2.3
Photo: Adam Verity.

Meeting places

A meeting of minds provides a chance for those involved in the search for or development of an idea to review the situation and engage in beneficial conversation. It is also an opportunity to talk through an idea in an attempt to gauge opinion and encourage feedback.

A particular issue might guide such meetings, although the process may actually be informal or adopt a freestyle approach.

The important thing is discussion: to understand the views of others, to understand need, and to explore potential directions. Gathering in spaces that are relaxed and free from pressure encourages open conversation.

Leading US industrial designer Tucker Viemeister has worked with many of the world's leading companies including Apple, Coca-Cola, Motorola, and Nike, and has been instrumental in the development of many leading US design offices.

Tucker explains the significance of the informal meeting place to develop ideas:

"I have been collaborating with Henry Myerberg for over twenty years and we've just bumped it up a notch in our new penthouse office. Like his library work, it is designed to encourage conversation and learning. Our 'office' space includes social gathering places and of course a kitchen, because that is a natural gathering place. When people work together they form a community and share their ideas freely—especially over a meal!"

Figure 2.4
"Every day (when the weather is not too bad) we eat together on our deck—is this work or play? It's certainly productive!"
Tucker Viemeister, President of Viemeister Industries, New York City, USA, 2014

Challenge understanding

To do the opposite of what is expected, to deliberately invite curiosity and debate, to pose a question about why something is the way it is and not some other way, is to instigate change.

The varied outputs of artists give an indication of questions that could or should be posed, questions that many dare not ask. Such questions are usually good questions, and although the context may need to change for them to be applied elsewhere, they remain good questions.

Time is needed to stop and think, to seriously think about what might be.

A tunnel-vision approach that constantly ignores everything around it, including the practices of others, is not a blueprint for success and will not introduce or stimulate different directions.

Figure 2.5
Jac Leirner
Little Light 3
2005
Electric cable, bulb, nails and socket
17 × 204 × 2½ in. (43.2 × 518.1 x× 6.3 cm)
© Jac Leirner
Photo: Courtesy of Galeria Fortes Vilaça and White Cube.

Figure 2.6
Mona Hatoum
Undercurrent (red) 2008
Cloth covered electric cable, light bulbs, dimmer device
3 1/8 × 314 15/16 × 314 15/16 in. (8 × 800 × 800 cm)
© Mona Hatoum
Photo: Murat Germen.
Courtesy of Arter, Istanbul and White Cube.

In the work of Mona Hatoum, Undercurrent (red) (2008), a stereotypical approach is challenged when the usually restrictive umbilical cord that tethers a light to a wall becomes central to the work in its freedom. The cable is also replicated many times to such an extent that it dominates the area and breathes life into multiple lights. Jac Leirner's Little Light 3 also makes the cable a key factor, and although potentially the light could be free, it remains attached to a wall, and irrespective of freedom the cable again takes center stage.

Understanding

The works of international artists such as Vladimir Rachev can become inspirational triggers for thought if viewed with an open mind, and if thoughts are allowed to flow.

Visits to galleries and exhibitions, and connecting with fine artists and sculptors, can influence the idea-generation process significantly. The designer needs to observe such works and aim to gain an appreciation of what is being communicated, and perhaps the fundamental scenario that prompted the works from the artist. Understanding such approaches can lead to new works, which might be quite different from the original inspiration, but which still retain some of the poignant characteristics that were initially intriguing.

Vladimir Rachev's Unfoldable Chair might prompt many ideas to the curious mind, but not necessarily to do with chairs. It is often the lateral connections that can be drivers for new ideas, rather than any literal interpretation.

It is not always important to visit a gallery or museum that is related to a particular area of interest, as so much information can be sourced from works in all fields if considered closely.

Figure 2.7
Vladimir Rachev, Unfoldable Chair.
Photo: Vladimir Rachev.

The unexpected, the unconventional, and the unusual form the basis for interesting outputs. The unexpected does not need to be too different from what is conventionally understood, but slight changes, such as a foldable chair that doesn't fold, or a series of chairs attached in an unfamiliar arrangement, do surprise. An understanding of habits and how objects are actually used rather than how they might be perceived invites alternative thinking and outcomes. Such outcomes break down boundaries and openly question sterile conventions. Designers Yvonne Fehling and Jennie Peiz create innovative works that are appealing due to their familiarity, but that are also very different to individuals' expectations of an object. Stuhlhockerbank is a work that the observer understands instantly through its familiar narrative, and yet conversely it is also an unfamiliar object.

In Stuhlhockerbank the chairs appear to break away from a bench into familiar seating clusters but unexpectedly remain connected to the bench structure. The Berlino Bench (2004) by Martino Gamper and Rainer Spehl, constructed out of various reclaimed seating components, adopted a similar narrative.

The Berlino Bench, however, used a variety of different seats including stools, chairs, and benches in the completed arrangement. It manages to capture many different stories within the work due to the eclectic nature of the sourced items.

Figure 2.8
Yvonne Fehling and Jennie Peiz, Stuhlhockerbank (Chairstoolbench).

Profiling

Profiling is an important and necessary investigative instrument that provides insight into the behavior, habits, and routines of individuals. As a design tool the process is not intended to be intrusive and can be conducted using primary or **secondary research** methods. The objective is to build an understanding—to probe where necessary, gathering detail to formulate patterns that can become useful directional signs.

What are the products that targeted audiences are purchasing? Where, how, and why do they use them? These are some of the issues that can be addressed using profiling. Collecting artifacts, images, materials, and other tangible evidence, which is believed to be associated with a particular group, begins to provide an understanding of what they like—strands of evidence that inform thinking.

Exploring and examining a range of sourced items can highlight commonalities that particular individuals are attracted to, such as an attention to detail and quality, but it can also depict issues such as financial positioning and brand awareness.

Target audiences are not simply related to areas such as age or gender, so **profiles** may be applicable to a range of individuals who might initially appear unrelated.

Figure 2.9
Design staff at the San Francisco based Astro Studios collate specific visual references to assist in identifying possible user scenarios and aspirations. Photo: Courtesy of Astro Studios.

Themes

> "In an age of abundance, design should be about formalizing values and statements. Every product should tell a story that hasn't been told before. Only this way can the process of designing something make sense and bring something to our world instead of burden it."
> **Eric Morel, Eric Morel Design**

Character adoption

Themes are necessary in the development of ideas and need to be carefully refined as an idea progresses. A theme usually has a relationship to the product being developed in some sense, although since this might be exploited through a single characteristic rather than a holistic literal interpretation, a theme may initially appear to be random or abstract to an observer. The theme may also be a reflection of a current trend, and therefore remote to the functions of the product, but appealing to a particular audience. The delightful Playmobilia stools by designer Tania da Cruz are playful, adopting a theme of toy hairstyles and colorful tones, but remain as viable, contemporary products due to the control of the work and the standard of execution. Achieving such an outcome as the Playmobilia stools might appear simple to the uninitiated or naive, but considerable thought needs to be directed at the subtleties and sensitivities of the work to ensure such purity and quality.

2.10

Figure 2.10
Tania da Cruz, Playmobilia stools. Exhibited at "SaloneSatellite," Milan, 2014.
Photo: Manuel Rio Casali.

Themes

The reference to animals in the j-me toothbrush holders transforms a mundane object into a vibrant and joyful object.

The adoption of a character in the development of an idea can be both fun and rewarding. It can give a sense of meaning and drive. Careful examination of the key characteristics of **animate** and **inanimate** objects can be exaggerated, simplified, and modified to influence idea progression, which can impact on emerging designs with great effect.

There doesn't need to be any obvious logic behind the embracing of a character, providing that appropriate design considerations and judgments are made, as outcomes are often open to individual interpretation.

Figure 2.11a and 2.11b
j-me, Bella and Grace toothbrush holders.

2.11a

2.11b

Why be conformist when something more exciting presents itself?

The alteration of a conventional form into a non-conventional form opens up an array of opportunities for how an object might be received. The Still Lives works by Fehling and Peiz, a series of life-sized pigs that are upholstered in a traditional fashion, are reminiscent of the zoomorphic Little Crawly Thing designed by Carl Clerkin, where upholstered stools had multiple cabriole legs added to suggest a scurrying bug.

According to Fehling and Peiz the Still Lives works can be received in many ways:

"Maybe these still lives simply eagerly strain the classic concept of design. In any case, they deliberately defy all definition. There is both something confounding and liberating about it. They are a welcome change in an unexpected direction."

In searching for an idea, it might simply be something unexpected or a break from the conventional that provides a valuable direction. The generation of ideas needs to understand the works of the past but does not need to adhere to previous conventions and attitudes. Still Lives are fun, useful, sculptural, appealing, and different. The execution of the work, however, remains at a very high standard and visual connections are made with more conventional works. The Still Lives works are a seating solution but also objects that prompt discussion and interest. Still Lives demand attention.

Figure 2.12
Yvonne Fehling and Jennie Peiz, Still Lives, Objects for Domestic Space. Photo: Frederik Busch.

> "The basic purpose was to design a rocking chair having no typical curved runners. I want to find a solution, which turns this well-worked principle upside down. So I used legs—usually standing for a solid stableness on the ground. Only the amount of used legs and their different lengths caused the same function like a runner. The contradictoriness makes Walker a unique object with the side-effect of looking like an animated being in motion."
> **Oliver Schick**

The generation of an idea may come from a simple desire to do something different, even if an existing solution is broadly acceptable. In looking for alternative solutions to accepted conventions the beautiful can emerge.

Designer Oliver Schick continually creates beautiful and appealing works from questioning his surroundings and observing what is actually taking place as a form of innovative reference.

The conceptual Walker chair designed by Oliver Schick manages to capture the essence of a traditional rocking chair and is synonymous with the Giacomo Balla painting at the Albright-Knox Art Gallery, Buffalo, New York, *Dynamism of a Dog on a Leash* (1912), where the painting captures the rapid movement of the dog's legs as it walks with its owner.

2.13

Figure 2.13
Oliver Schick, Walker, 2007.
Photo: Michael Himpel.

Cultural references

The term "cultural" is somewhat ambiguous and is often subjected to different interpretations, which can relate to objects as well as to social interactions and behavior.

It is important to allow yourself to encounter different cultural experiences and to view different approaches to living with an open and receptive mind. Ideas can be sourced through an awareness of different cultures and cultural habits.

Being unaware of the ways in which different cultures interact and respond to situations can be a mistake.

Anthropologists study human beings from a broad range of perspectives, including their customs, faith, and culture, as well as physical issues and situations. To appreciate cultural influences it is necessary to watch, listen, ask, understand context, and comprehend reasoning. Approaches such as the Stool (Guan Ziyin, 2014) and Play Me I'm Yours (Luke Jerram, 2008) provide a good foundation for cultural referencing (see Chapter 1).

Culture might be defined as a way of life, and can relate to localized organizations and groups as much as to the practices of individuals from different countries or backgrounds.
Influences should be gathered from as many diverse cultural experiences as possible. Constraints can compound situations where information needs to be gathered. However, irrespective of such pressures, observation of practices, cultures, and methods must be given sufficient time to ensure that interpretation is accurate and that assumptions are not being made. It may be necessary to interact with a community, organization or social group for a reasonable period of time. Interaction will provide opportunities to acknowledge procedures, practices, and approaches to a broad array of tasks, some of which may be remote or contrary to familiar practices and habits. The diverse ways in which issues are addressed in a different cultural environment are very necessary aspects of the design process.

It is important to recognize that an accepted practice in a certain culture may be deemed as an insult or be frowned upon in another. Better understanding of different cultures and practices will almost certainly improve creativity in design. Observing the ingenuity, inventiveness, and imagination of cultures with sparse resources can be as beneficial as observing more developed cultures.

> "The National Palace Museum is the Louvre of the East."
> **Alberto Alessi**

Collaboration between Alessi and the National Palace Museum in Taiwan enabled Western and Eastern cultures to merge and an understanding of Chinese traditional practices to be influential in the creation of the Stefano Giovannoni pieces Mr Chin (2007) and OrienTales (2008).

Understanding cultures, exploring historical references, and appreciating details and applications influence design direction and thinking. Stefano Giovannoni combined traditional practice of the East observed through observing the craft of the East.

The prestigious National Palace Museum is the custodian of many significant examples of art and culture from the Ch'ing, Sung, Yuan, and Ming dynasties of ancient China. The works demonstrate a meticulous attention to detail and craftsmanship in beautiful materials including jade, bronze, and porcelain.

Figure 2.14
Giovannoni Design, Alessi S.P.A., OrienTales.

The Alessi works are created as if they were ancient artefacts: crafted, painted, and assembled in a manner similar to methods for creating wooden or porcelain works. The background referencing is detailed and thorough and has an honesty connected to its origins, but also manages to merge the historical importance of cultural works with the unique Alessi style: two different cultures, steered by meanings and merging through attention to detail and craftsmanship.

Alessi describes the care, attention, and direction given to their works, influenced by traditional Eastern practice as: "Eastern stories through Western eyes."

It is important to engage with different cultures and practices, to open up to opportunities, and to acknowledge the diversity of customs that can inform creative directions.

Figure 2.15
Giovannoni Design, Alessi S.P.A., OrienTales: The Banana Family characters.

Historical references

Not to investigate or consider previous approaches to design, alternative philosophies, constraints, and restrictions is to miss an opportunity to explore a wealth of exciting ideas that may have been forgotten. Many historical influences still have much to offer, although the original context changes.

There is a need to question why successful products ceased to exist, and to understand the external pressures that may have been significant in their demise.

Ideas are too often lost to history for a broad range of reasons, but undoubtedly some could be re-evaluated and resurrected, either in their entirety or just a particular facet.

Fluctuating trends, cultural differences, competition, technological advancements, or a loss of skills are all significant factors that can bring about change. It is unlikely that all the components of a historical idea are irrelevant to the contemporary consumer.

Attention to detail, quality of craftsmanship, and material usage are among many of the historical agendas that remain important, and any such references should be seized.

Design historian Renny Ramakers and designer Gijs Bakker co-founded the globally inspirational Droog Design in 1993 and in 2011 Renny Ramakers founded Studio Droog.

The Family Vase collection (2013) created by Studio Droog as part of The New Original project demonstrates the ability to take inspiration from a historical base—in this case, the Famille vases of seventeenth and eighteenth-century China—but also looks to respond to contemporary issues and the reputation of a Chinese copycat culture.

Conducted in Shenzhen, Studio Droog looked to copy the copycat, but in a manner that would demonstrate how an alteration in thinking could inspire originality. The vases in the collection represent a stereotypical form from the seventeenth- and eighteenth-century period, but rather than attempt to copy the traditional patterns on the vases, color was replicated in representative bands. The Family Vase collections are contemporary originals inspired by a copycat culture.

Figure 2.16
Studio Droog, The New Original collection: Family Vase, 2013. Historical Chinese vases inspired the Family Vase collection by Studio Droog.
Photo: Mo Schalkx.

Themes

The Studio Droog Black Family Vase in the New Original collection is created using rapid prototyping—a process removed from and incongruous to the original makers—but retains an inherent honesty. The various color bands that make up the vase are proportional representations of the original color used in the distinctive artworks that exquisitely and sensitively adorned the original seventeenth- and eighteenth-century Chinese vases. The form is a generic shape, a seemingly distinctive structure (although different), but one still associates it immediately with the historical wares of the period. The approach rejects a direct copy culture in preference for what might be considered a reconfiguration, and in doing so it creates a viable contemporary descendent of the originals.

Figure 2.17
Studio Droog, The New Original collection: Family Vase, 2013.
Photo: Mo Schalkx.

"The Sony Walkman II entered the market in the late seventies. At the time I was very excited about the device's extremely small size. Not a millimetre was wasted. The standard audio cassette was in effect encased in a thin metal skin containing all the components, such as the drive, audio head, and battery.

"The thing actually felt no larger than the cassette itself. Impossible to do, right? I tried anyway! Why enclose the cassette? I left it open, showed it, and only created a form-fitting connection with three mechanical openings. The drive, audio head, and battery subsequently hang as a visually separate unit alongside the cassette. Although the resulting Museman had a somewhat larger volume than the Walkman II, the technique of splitting the housing into two separate entities made it appear smaller."
Jörg Ratzlaff

The Museman designed by Jörg Ratzlaff for frog design exemplifies an alternative thinking approach. At a time when the Sony Walkman was the predominant personal stereo and a great deal of focus was on the physical size of the product, the proposed Museman took the innovative step to question the need to place a box (the cassette) inside another box (the personal stereo). The outcome was a design that attached onto a cassette rather than surrounding it.

Figure 2.18
Jörg Ratzlaff, frog design, Museman.
Photo: V. Goico (Image Museman).

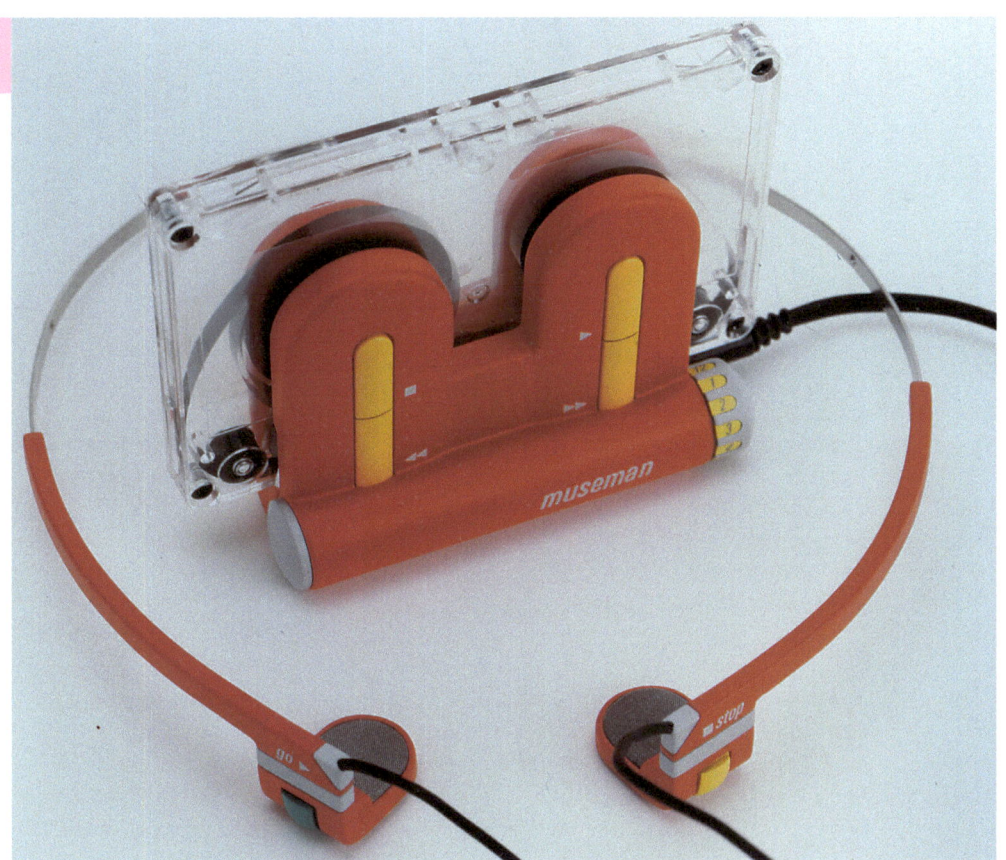

86

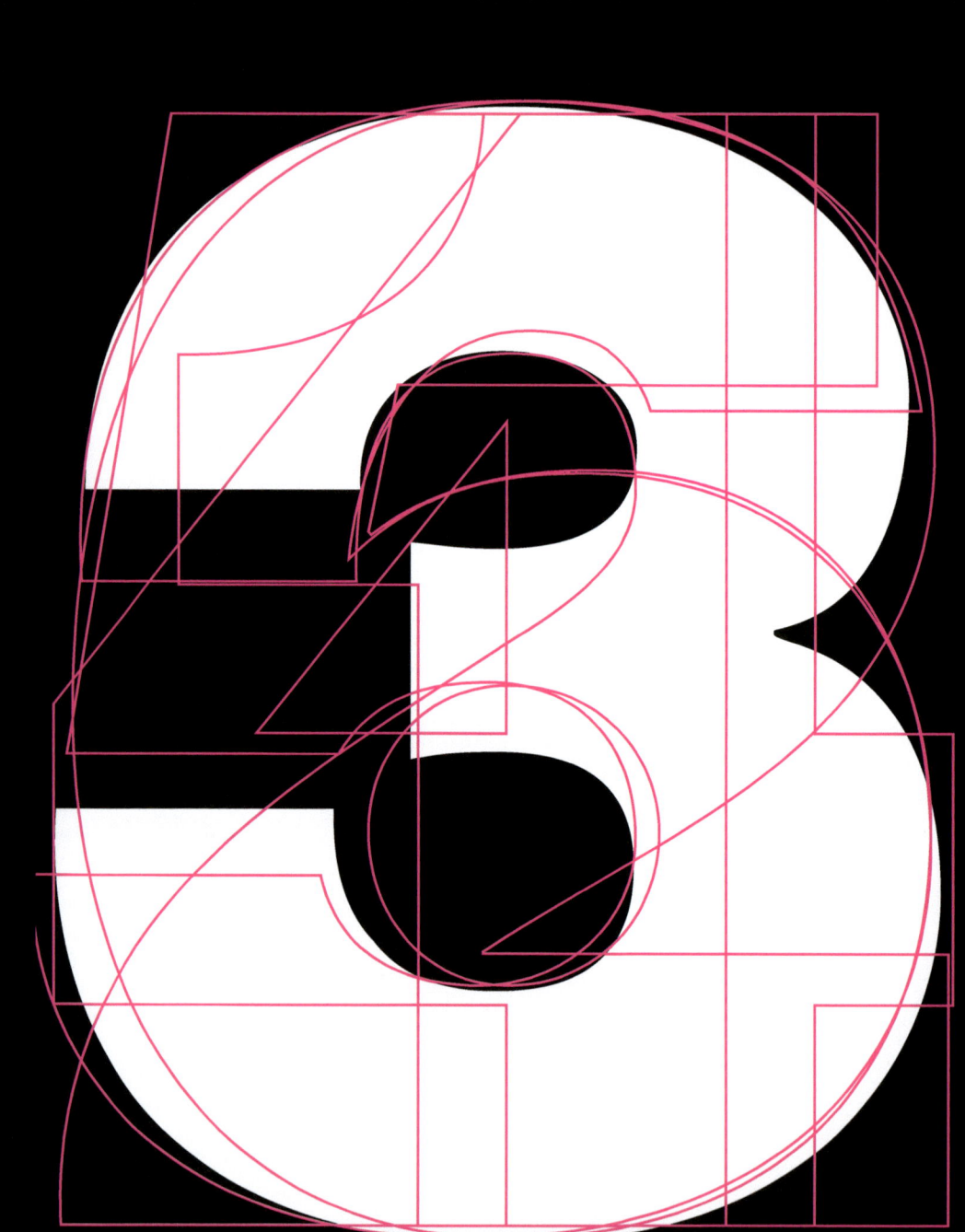

Experimental Beauty

Looking beyond the familiar and the comprehensible, to experiment or simply play with material characteristics provides a platform for original thought where attitudes and hackneyed values can be steered toward a particular creative path. Sometimes controversial, sometimes abrupt, the desire to create the unfamiliar is often prompted by embracing and exploring the cultures and values of disparate disciplines and ultimately gaining valuable insights that inform.

Materials

> "When I am working on a problem, I never think about beauty ... but when I have finished, if the solution is not beautiful, I know it is wrong."
> **Buckminster Fuller**

Contrasts

Something that is different naturally engages the imagination and stimulates curiosity. It often evolves from experimental thinking and invariably sits comfortably with high creativity. Such an approach is usually addictive and captivating, and enables beauty to be introduced.

The natural instinct of experimental practice is to inquire further, to want more answers, and to become increasingly submerged in associated values.

When there is no opportunity to experiment it can become difficult to sustain motivation or remain inspired. It is important to ensure that creative activity does not evolve into a burden, since all initial activities should be creative and be allowed to drive the search for ideas.

The instigation for a thought can come from all directions, and inspiration from personal discovery should be allowed to feed into the creative process.

The more interesting the references to work with, the greater the probability of achieving a beautiful outcome.

A "why not" or "do what others don't" attitude provides the creative platform to legitimately deviate from what is expected. Breaking new ground might relate to ideas, but it may also mean new materials, or a standard of quality or detail that is unexpected.

A desire to do something alternative and maybe controversial will generate original thinking and shake up practices that have become too habitual and comfortable. The execution and attention to detail of the Barber & Osgerby limited edition Murano glass vases for Venini exemplify craftsmanship and set a material benchmark.

The nautical-inspired glass vase within its protective cage is unusual, beautiful, and original, an outcome achieved through skill sets coming together, daring to be different, and sharing experience and knowledge.

Figure 3.1
Edward Barber and Jay Osgerby, Limited edition Lanterne Marine – Murano glass, for Venini (2009).

3.1

> "I distinguish between laboratory and factory."
> **Maria Kirk Mikkelsen**

The range of materials that are available to the designer is fantastic. Inspiration for working with materials can be sourced from different disciplines and cultures, which provide stimulating possibilities and sources of inspiration.

Materials should be valued, challenged, and pushed to their limitations—limits that would not normally be considered possible. Investigating a material and exploiting its beauty through craftsmanship can realign preconceived perceptions of what is possible.

Designing with materials can take different routes. It is possible to conceive an idea and then find an appropriate material to function in a specific role; it is also feasible to explore a material and then adopt a use for it; and of course it is always possible to simply play with a material and then go and do something completely different.

Materials contribute to the soul of a design and their physical and mental beauty should be appreciated, respected, and enjoyed.

Maria Kirk Mikkelsen's description of a "laboratory" approach refers to a range of creative thinking activities and investigations, the development of experiences, stories, and ideas, whereas the "factory" approach is informed by "laboratory," and is regarded as being more material-based, instinctive, and expressional.

The visual language of a product can be manipulated to emphasize a diverse range of signals that may be consciously or subconsciously recognized by a third party.

It is not always necessary to consider the semiology of a product in detail, as the message it communicates may simply be a culmination of other considerations, but in some cases the signals portrayed are fundamental to success and need to be recognized.

Attributes associated with a product such as form and materials can be selected to enhance a desired or inherent message. The selections might be basic but they can be effective communicators if considered carefully. Many products are categorized or stereotyped due to their particular visual language, but considering an alternative or unfamiliar language can refresh old perceptions.

> "There are no rules here—we're trying to accomplish something."
> **Thomas A. Edison**

Rules are there to be broken. In the 1960s, Anti-Design groups including Archigram (1961), Archizoom (1966), and SUPERSTUDIO (1966) challenged accepted values and in particular the notion of "form follows function."

Taboos were embraced—something that became less shocking and more mainstream. Accepted design rules were challenged, questioned, and overturned.

Designer Ettore Sottsass founded the Italian design group Memphis in 1980, which again questioned contemporary approaches and asked "What is design?" The group reintroduced color into design, generated asymmetrical outputs, and merged material extremes in an attempt to distance themselves from the bland design that had become prevalent. The designs of Ettore Sottsass and the Memphis group ultimately influenced a generation with their innovative thinking and questioning approach.

Why are things done the way they are? Is it too dangerous to depart from the **comfort zone**? Exploring and systematically interrogating the various components of a product or the conventional methods of manufacture can conjure up a multitude of potential directions, as well as querying the purpose of accepted approaches or practices. An audience seldom realizes what is required until it is presented, as premature judgments relate only to previous encounters, knowledge, expectations, and experiences. When a method of doing something is called into question, it is likely to unleash a myriad of exhilarating proposals that challenge values, and a "will do" attitude.

What works, what doesn't? Is it possible to ascertain an immediate outcome? Experimenting, investigating, exploring, and inquiring allow for ideas to be teased out. Making mistakes, incorrect judgments, and questioning is all part of the experimentation strategy. Making a mistake can often be a positive thing, and an opportunity to reject a particular line of inquiry. A mistake may even enlighten and provide previously unconsidered opportunities. Approaching a problem expecting to make a mistake while exploring ideas, especially those that might merge disciplines, provides confidence and an ability to progress more easily than working in an arena of constant caution and anxiety.

Exploration

> "We are testing the future, not predicting it."
> **Jack Mama, former Creative Director, Philips Design Probe**

Experimentation

A crossover of practice and experimental endeavor can present many diverse and beneficial indicators in the search for a creative direction.

Practitioners who are able to venture between art and science or technology continually set creative benchmarks and provide material that can be both inspirational and constructive to others. Experimentation and an ability to communicate findings to a wider audience create dialogs of possibility that can prompt further questions and more intense activity.

The Bio-light, a harbor of fluorescing microbes in a **chemostat** of bespoke labware, immediately catches the imagination and suggests potential opportunity.

Microbial Home is a probe, a far-future design concept. It is not intended as a production prototype, nor will it be sold as a Philips product.

Figure 3.2
Philips, Microbial Home Probe, Bio-light.

> "The Microbial Home Probe project consists of a domestic ecosystem that challenges conventional design solutions."
> **Miep Swaminathan, Philips**

The Microbial Home Probe project bridged creative disciplines, the scientific, and the artistic, and in doing so asked significant and extensive questions to reveal what might become possible. The outcomes are different but familiar, embracing the wonders of the animate with the beauty of the inanimate, the purity of the synthetic with the elegance of the natural, and visionary practice with contemporary practice.

The Microbial Home Probe questions conventional thinking and in doing so recognizes that the output of one function has the ability to become the input of another function—an idea that echoes the importance of the designer exploring the outputs of others to inspire their own lateral directions.

It is increasingly important in a search for the designer to identify original paths, to venture beyond what they know, and gain understanding of other disciplines through collaborating with those with specialist knowledge. Tim Brown at IDEO suggests that designers need to adopt a T-bar approach to design: a situation where they are competent in a particular field but are also aware of their surroundings.

The ability to facilitate cross-disciplinary discussion and to broker experimental and engaging activity between science and art is a characteristic of the internationally acclaimed C-Lab. The research of C-Lab, co-founded by synthetic and biological artists Howard Boland and Laura Ciniti, explores scientific and artistic contexts where the relationships between animate and inanimate ideas are able to mature and become sophisticated entities that inspire others.

The Philips Bio-light and the C-Lab installation Stress-o-stat present inspirational insights for other creative thinkers, which would not be available if the initiators had not asked poignant questions. The setups—chemostats—are able to monitor and adjust the conditions of bacteria implemented with a fluorescing protein. An exposure to stress triggers the bacteria in the setup to express the protein as light.

> "Stress-o-stat is a result of an immersive and independent laboratory practice using synthetic biology to develop new types of artistic expressions. A special genetic switch involved in stress response was located and combined into a genetic construct to produce fluorescing proteins. Once implemented in bacteria, fluorescing proteins are expressed during oxidative stress producing a yellow-green color in response to blue light."
> **Howard Boland, synthetic biological artist and co-founder of C-LAB**

Researchers such as Donna Franklin have also explored biological activity in the creation of garments, and, in conjunction with Garry Cass, created the Micro "be" collection through the creation of cellulose from a colony of *Acetobacter* bacteria.

Researcher, designer, and microbiologist Suzanne Lee combined scientific and design disciplines to create her Biocouture collections. Although a fashion designer, the practice of merging disciplines is applicable to all the creative areas, and the outputs of Suzanne Lee, Donna Franklin, and others provide a realistic incentive in the search for inspiration.

Figure 3.3
C-Lab, Stress-o-stat, final setup, 2011. This work uses a chemostat to maintain constant cell population using a feed (left), a bioreactor attached to a Graham condenser (center), and a deposit (right). Varying nutrients from rich to poor, cellular stress is seen as light emitted by proteins as a result of genetic engineering.
Photo: C-Lab.

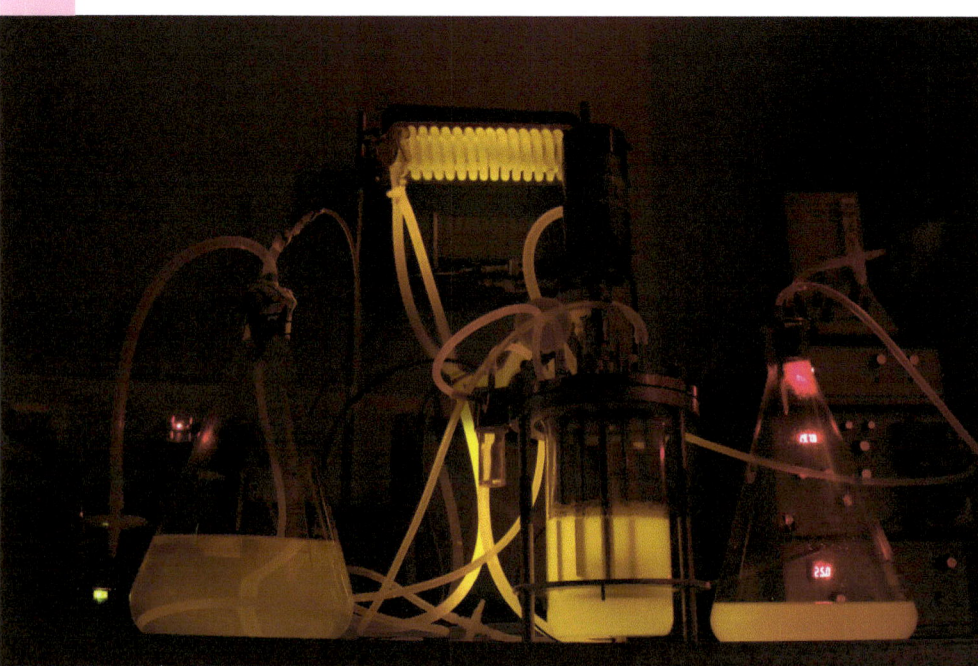

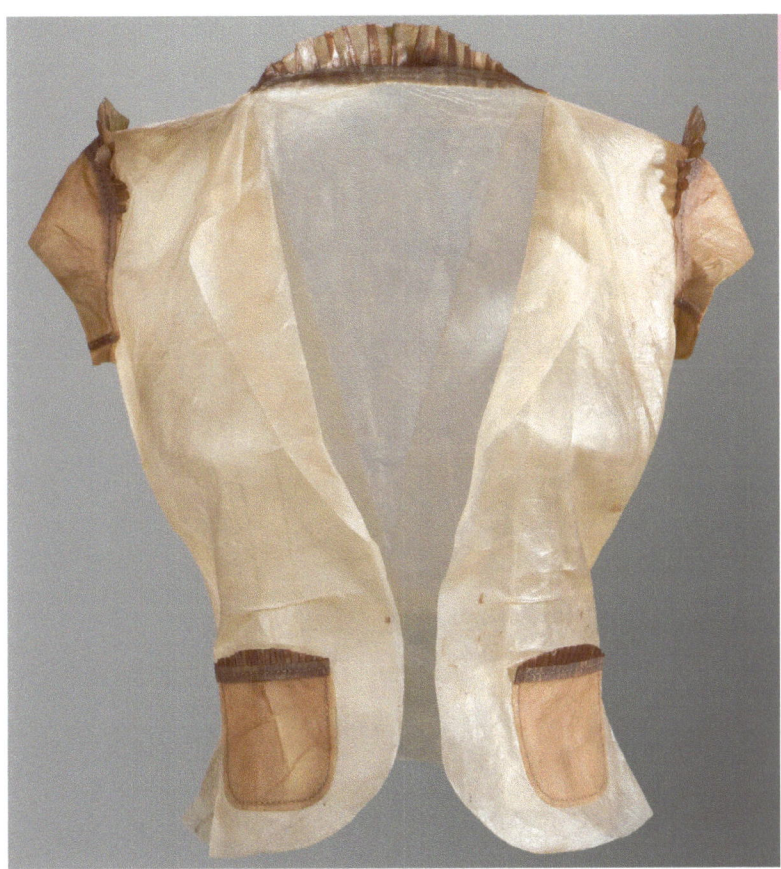

Figure 3.4
Suzanne Lee, Biocouture. Curiosity and a fusion of minds coupled with experimental play enabled Suzanne Lee to create her innovative Biocouture collection. Bacteria created the material for the garments.
Photo: Science Museum / Science & Society Picture Library

Exploration

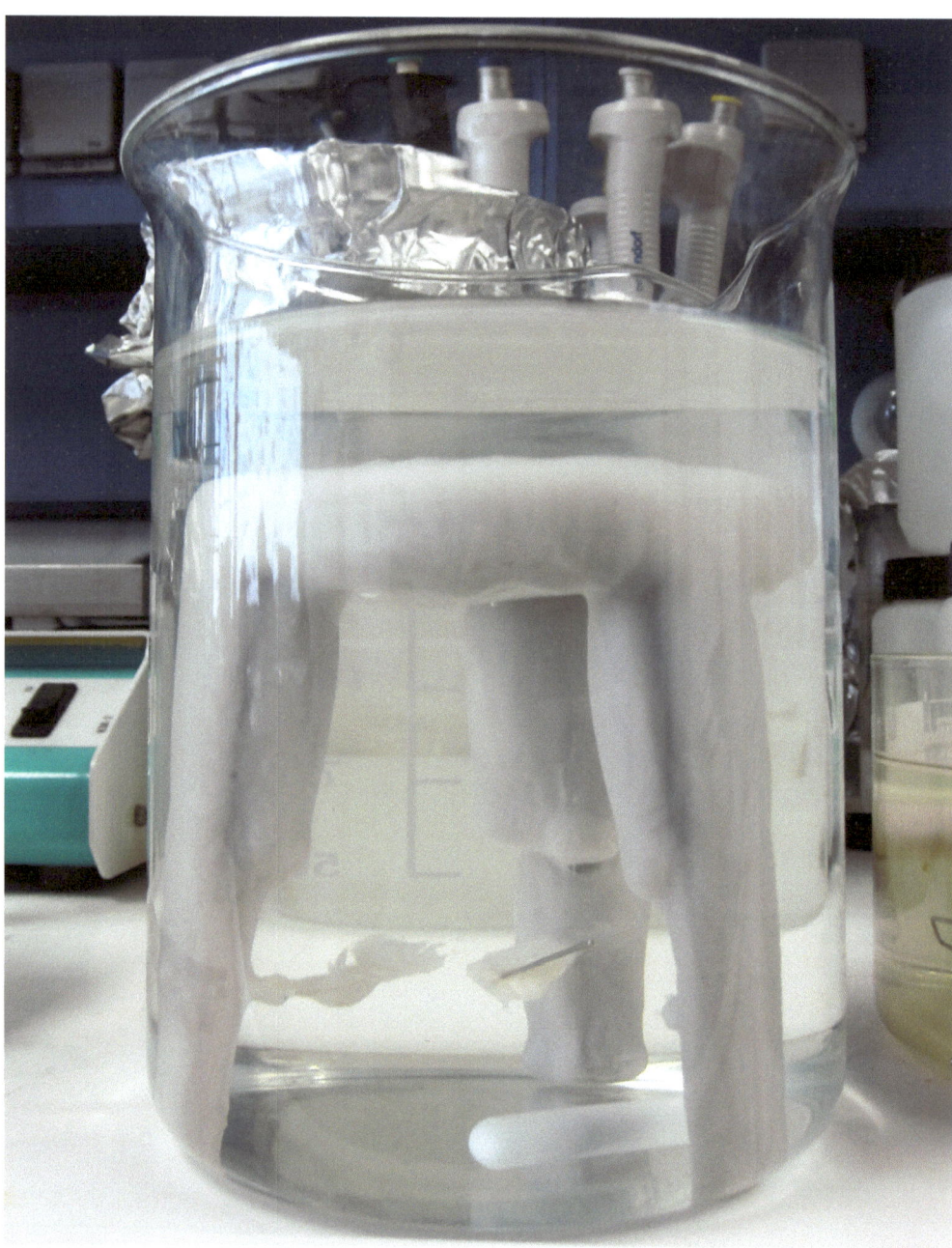

Figure 3.5
Jannis Huelsen, Xylinum stool. A cellulose film gradually surrounds a submerged wooden stool, in a flask of microbial activity.

Exploring the practices of others and appreciating cross-disciplinary activity—activities outside of the creative arenas associated with the arts—introduces alternative directions to creative thinking.

Jannis Huelsen recognized the properties of microbes to create cellulose, submerging a wooden stool, a form of biological scaffold, in a concentrated broth to allow *Acetobacter xylinum* microbes to cover the stool with a shroud of extracellular material.

The viscous substance covering the predetermined scaffold is dried in a similar fashion to the garments made by Suzanne Lee, but unlike the garments, the drying process adheres the microbial matter to the frame.

The use of biological practices within design is burgeoning and opportunities for cross-disciplinary dialog should be encouraged as both parties have information to share.

The curious, dried coverings of the Jannis Huelsen stools are unusual, but perhaps only because they are unfamiliar. Unknown territory is where a creative mind should aim to be active.

The IDEO design studio in California has also explored the potential of microbes and a form of natural, additive fabrication with the work of scientists, where stimulated *Escherichia coli (E. coli)* creates a solid form of cellulose for the generation of simple products.

3.6

Figure 3.6
Jannis Huelsen, Xylinum stool. The cellulose film attached to the Xylinum stool dries and reduces when removed from the flask of microbial activity.

The materials available to the designer are not simply limited to original, untouched virgin materials, since if a material can be worked it immediately becomes a viable proposition, irrespective of whether it already has a particular form. Designers have a responsibility to explore all options in their creative journeys, and considering materials is fundamental. A material in a preconceived form of any description presents a challenge to the designer, but ultimately it is the ability to look beyond the obvious and consider things differently that remains central to the success of an outcome. It is always necessary to ask "What if?"

Designers Tejo Remy and René Veenhuizen regularly challenge materials and consistently demonstrate an ability to see opportunities and inspiration for their creative outputs in everyday objects. The use of familiar materials and objects in unexpected contexts creates a connection with an original story, but also provides an opportunity to make the transition to the surrogate narrative.

The playful and experimental approaches of Tejo Remy and René Veenhuizen provide a creative platform for imaginative applications that are distinctive and stimulating. Opportunities for the use of different materials continually evolve due to the practices of inquisitive minds and a compelling desire to try something original.

Figure 3.7
Tejo Remy and René Veenhuizen, Concrete Furniture, Soft Moulded Bench, 2012.
Photo: Ernst Moritz.

A desire to experiment and a curiosity about encouraging concrete to adopt a predetermined pattern led Tejo Remy and René Veenhuizen to tailor a series of flexible formers and bespoke frames to steer the seemingly unpredictable material. But they also anticipated an element of distortion and desirable uncertainty in the eventual outcome.

Their distinctive Concrete Furniture works create the impression of the material being inflated and initially challenge the preconceptions of concrete.

Figure 3.8
Tejo Remy and René Veenhuizen, Soft Concrete Bench, 2012.
Photo: Ernst Moritz.

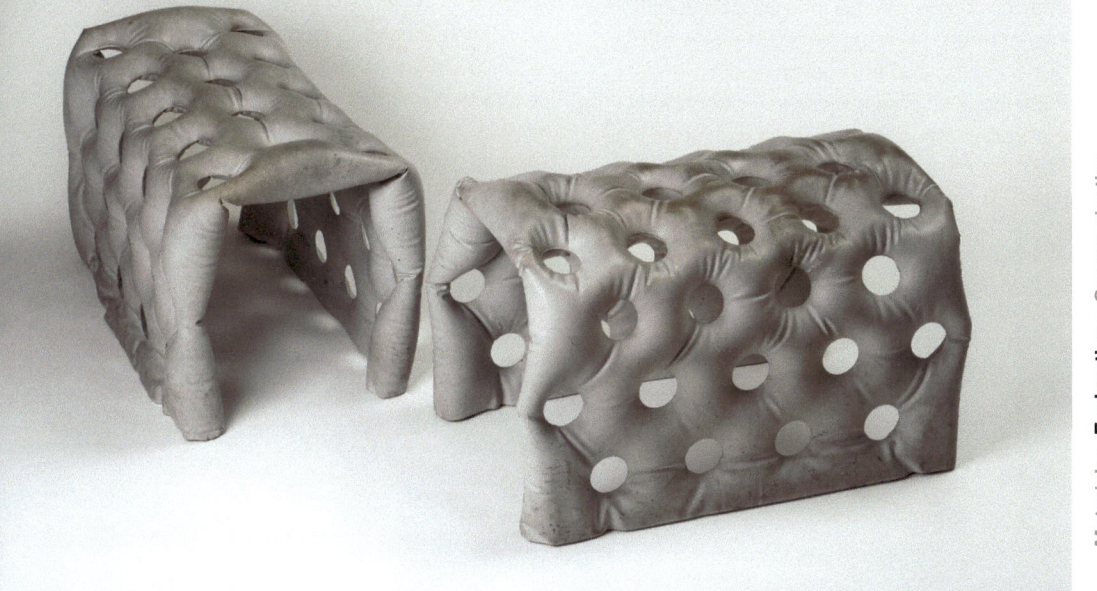

Play

An experimental, hybrid play approach to the generation of ideas can be used to arrive at beautifully bizarre and creatively diverse outcomes. Proposals that might normally be neglected or considered a high risk can be pursued and sensitively realized through a mix of ideas combining many references and disparate elements.

If a design process isn't productive it is necessary to ask why.

The blending of materials and a disregard for accepted rules and taboos invites the fantastic and the sublime to take center stage.

As previously mentioned, the Anti-Design studios of the 1960s daringly shook up conventional ideas and thinking, and in doing so inadvertently demonstrated that sometimes, in an attempt to be controversial, the outcome becomes surprisingly agreeable.

Where there are no boundaries to the imagination and where there is a willingness to be different, something wonderful can emerge; but when the imagination is stifled and starved, potential outcomes are at risk of becoming dull and complacent. A mind-set that challenges restrictions and embraces the unthinkable can alleviate frustrations and provide the necessary stimulus to reboot the imagination.

Green Chicken Rocking Horse and Rocking Sausage Hot Dog (2012) are some of the many creative outputs that were produced by the constantly burgeoning imagination of Jamie Hayon, the influential Spanish designer who demonstrates an ongoing ability to mix seemingly eclectic ideas and elements into beautiful outcomes. A fantastic and fearless disregard for restrictive, creative precedents relating to materials, scale, or theme, combined with an overwhelming adherence to detail and craftsmanship ensure that the outcomes, such as those exhibited at "Funtastico," are magical and have a capacity to momentarily transport audiences to another world.

It is the adventurous and audacious who ultimately create the adventurous and the audacious. Such leaders are courageous and confident in their own abilities, having experimented and explored many different directions to affirm their ideas.

It is such outcomes that stimulate the imaginations of others and that can move collective thinking forward.

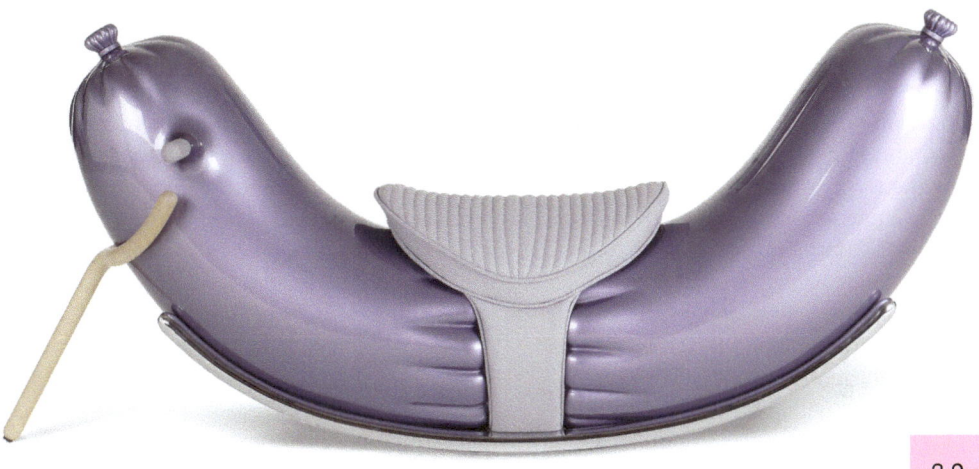

Figure 3.9
Jamie Hayon and Nienke Klunder, Rocking Sausage Hot Dog, 2012. Exhibited at "Funtastico," 2013–2014, a ten-year retrospective of the creative works of Jamie Hayon, Groninger Museum, the Netherlands.

Communication

Scale

Experimentation with scale presents a spectacular opportunity to seek out diverse references in distinct situations, and then translate any significant elements to a contrasting scale to elicit an alternative, unexpected narrative.

The control of scale can introduce hidden worlds. Previously unseen beauty and wonder can be openly captured and revealed through the repositioning of objects on an unfamiliar stage. The imagination is triggered through such unexpected changes.

In 1989, Dietmar Henneka photographed the Sun SPARC hardware stations designed for Sun Microsystems by frog design. The photograph, titled "Sunday," featured on the back cover of the July edition of Design magazine, depicted the various SPARC stations surrounded by miniature figures creating the impression of working and living in a major city.

Designer Dominic Wilcox uses similar customized figures set in miniature scenes in his Watch Sculptures works, such as Oblivious iPhone User, Hide 'n Seek, and the Sitting Man.

The theme of the Oblivious iPhone User exemplifies the importance of observing one's surroundings and of taking inspiration from everything.

Figure 3.10
Dominic Wilcox, Oblivious iPhone user.
Photo: Dominic Wilcox.

Figure 3.11
Dominic Wilcox, Hide 'n Seek.
Photo: Dominic Wilcox.

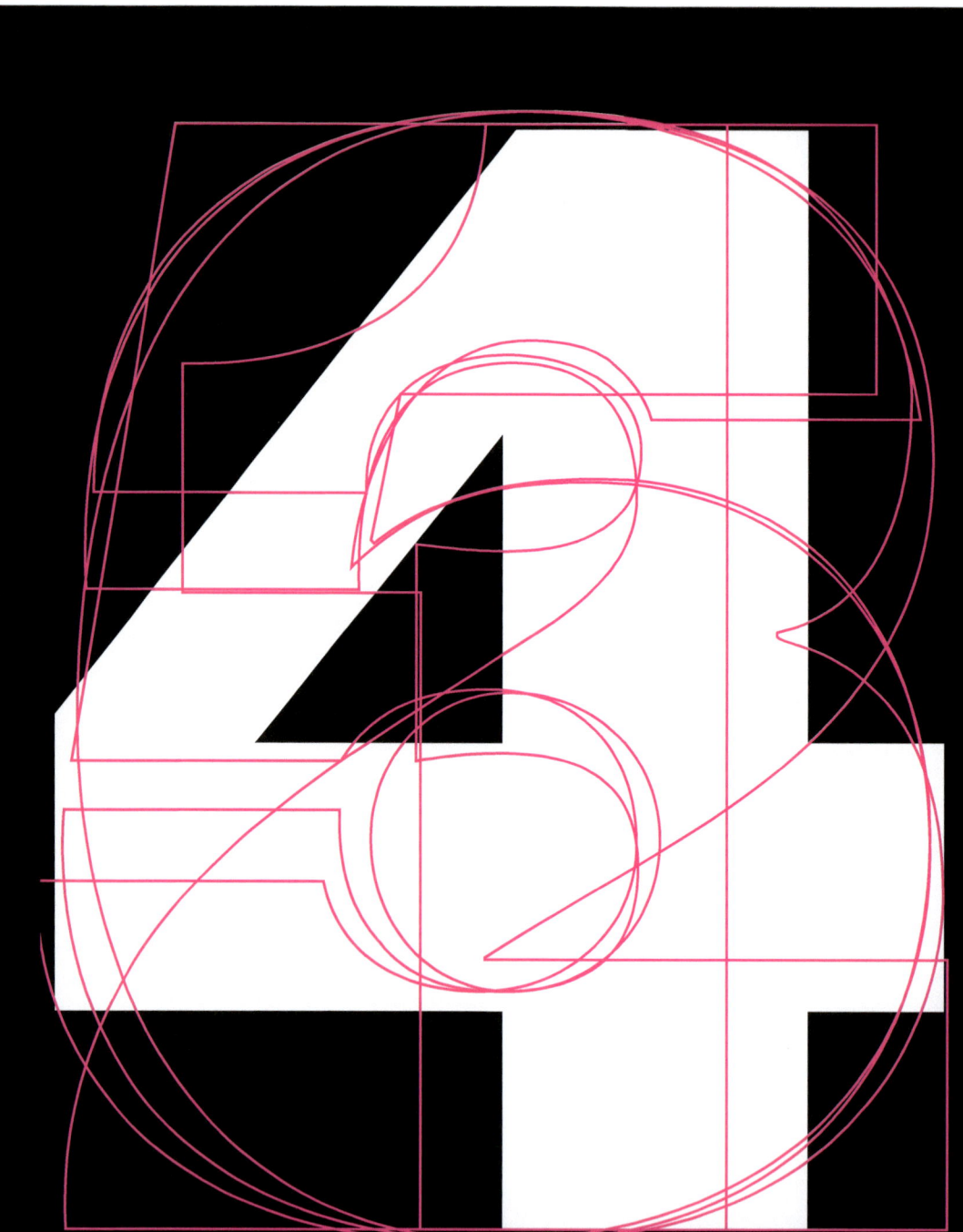

Sensory Perceptions

Subtle changes in design can have a significant impact on the way information is perceived, and a tendency to overwork a design can be detrimental to the outcome. Emotional responses that an individual, a society, or a particular culture might express can appear rational or irrational but they are responses that need to be recognized and accepted. In addition to the core and dominant aspects of a design, peripheral elements and added values can significantly influence perception and understanding.

Senses

All of the senses should be engaged when considering ideas so as to gain as much inspiration from the surrounding environment as possible. Inhibitions and inexperience should not prevent ideas from being explored and considered, but rather they should be a base from which to inquire and consider what can be communicated.

It is often the case that if a situation feels right—even if you haven't identified the particular aspect behind it—it should be explored and examined for possible adaptation or transposition. In such circumstances, the situation is perhaps more instinctive and visceral, and such occasions should not be overlooked. The spirit and soul of an idea is a combination of sensory and instinctive awareness.

It isn't necessary, or even always desirable, to try to replicate all of the characteristics of something that generates interest; but it is necessary to try and appreciate what particular facet of something is creating the interest, capturing the imagination and demanding further attention.

The senses are continually at work through conscious and subconscious means, and everyday experiences can become a creative source. The way the senses respond to different stimulations changes depending on external factors; therefore something might only be of interest in a particular setting, at a particular moment in time, or when presented under certain conditions to a particular individual. The aim is to translate the essence of what is being observed or understood through the different senses into a tangible outcome—an outcome that is able to retell or communicate a particular experience.

It is often very difficult to accurately convey a specific emotion or primary response to a broad audience due to the context being misunderstood or misrepresented by individual users.

Color

If someone is asked what their favorite color is, the response could be any number of things, but it is meaningless without a specific context or reference. Many colors can be considered within a particular genre, and each type communicates a different message.

Color is an important component in the design process, and if it is not considered carefully it can also become the deciding factor between a product being successful or not, no matter how well considered it is in all other regards.

Color is generally recognized before any other aspect of an object, including overall form, and so in the split second in which an individual makes a decision about something, the color makes a significant impression. The color may even be desired by the user, but can be readily dismissed due to the tone being misplaced.

It is often the case that colors are not considered in the initial stages of an idea emerging and frequently only make an appearance in the latter stages of the design journey. There is a difficult balance to strike in the use of color, but generally the sooner it is associated with an idea the better. The reason why color often makes a late appearance in the design process is perhaps understandable in the sense that a design or idea could be unfairly rejected in the initial stages simply because of a particular color association, but color should be considered to be a functional component.

Color theorists have often demonstrated that color can be associated with a range of characteristics due to previous tangential experiences, but such theories tend to be generic and are often too vague for meaningful appraisal in the design process.

Colors of all description can be considered to be heavy, light, strong, weak, soft, hard, clean, or dirty, in addition to many other characteristics. Inspiration for color selections and combinations are everywhere.

Figure 4.1
Beijing Bowl – Ms. Kitty.
Photo: Qicong Lin.

4.1

Color inspiration is considerable and it is often the subtleties in color that are of particular interest when making a selection.

In Beijing, China, it is not uncommon to see washing lines shrouding the outside of community tower blocks, or plastic basins lining the streets outside public and private buildings. These items are often sun-bleached due to being continuously exposed to sunlight. The variety of color tones within such simple everyday objects is fascinating, and provided the initial inspiration for a design at the Tsinghua University international foundation program in Beijing.

Figure 4.2
A sun-bleached plastic bowl on the Zhongguancun Donglu Road, Beijing, China, 2014. The color was the key inspiration for the design of the Beijing Bowl – Ms. Kitty lamp at Tsinghua, a design enhanced by the story connected to the bowl and by the various scratch marks on the outer surface of the plastic.

The Beijing bowl – Ms. Kitty lamp exploited the potential of cheap plastic products and their characteristic translucence. The lamp was the study that inspired the Guangzhou Firestick chandelier (2014), which was also made from cheap, redundant plastic.

The scratch marks on the bowl, a testament of everyday life and time spent on the street, are a different tone to the main body of the bowl, and when illuminated light highlights these materially reduced regions. It is perhaps unusual for a color to be the inspiration for an outcome, but it isn't unusual to discover inspirational material in the most unlikely of circumstances. Simply describing the Ms. Kitty lamp as pink would not sufficiently describe the colors of the lamp and the inherent warmth and strength of the design.

Figure 4.3
Liu Youlv, Yang Jun, Jia Xinyi and Chen Pengpeng, Beijing Bowl – Ms. Kitty lamp, 2014. Photo: Qicong Lin.

Texture

Holding, feeling, touching, caressing, and manipulating are activities that provide an understanding of a material or form that can trigger the imagination. The craftsman will often hold a material to feel and sense quality prior to working with it in a particular way; the way something feels is of the utmost importance.

The sensation of touch can be encountered anywhere, at any time, and the sudden realization that something feels fascinating should be noted.

Indicators such as sculptures or banisters that have been polished through generations touching, feeling or enjoying them might provide inspiration for a possible direction, as might the weathered texture of shells, wood, or rocks on a beach. Such delicate forms can be held and caressed to appreciate overall form, but can also further stimulate understanding of textures.

Textures can often be incorporated into a design to hide manufacturing defects as well as to enhance the property of a particular object. Seeing through touch on a continual basis will provide a valuable understanding of textures and their significance in the search for ideas.

It should be recognized that not all touch is conducted using the hand, and that it relates to the overall body experience.

4.4a

4.4b

Figures 4.4a and 4.4b
The Buddha Maitreya sculpture at Lingshan Wonderland, Longshan Mountain, Wuxi, Jiangsu Province, China. The prestigious bronze sculptures exhibit a distinctive patina created through continual touch.

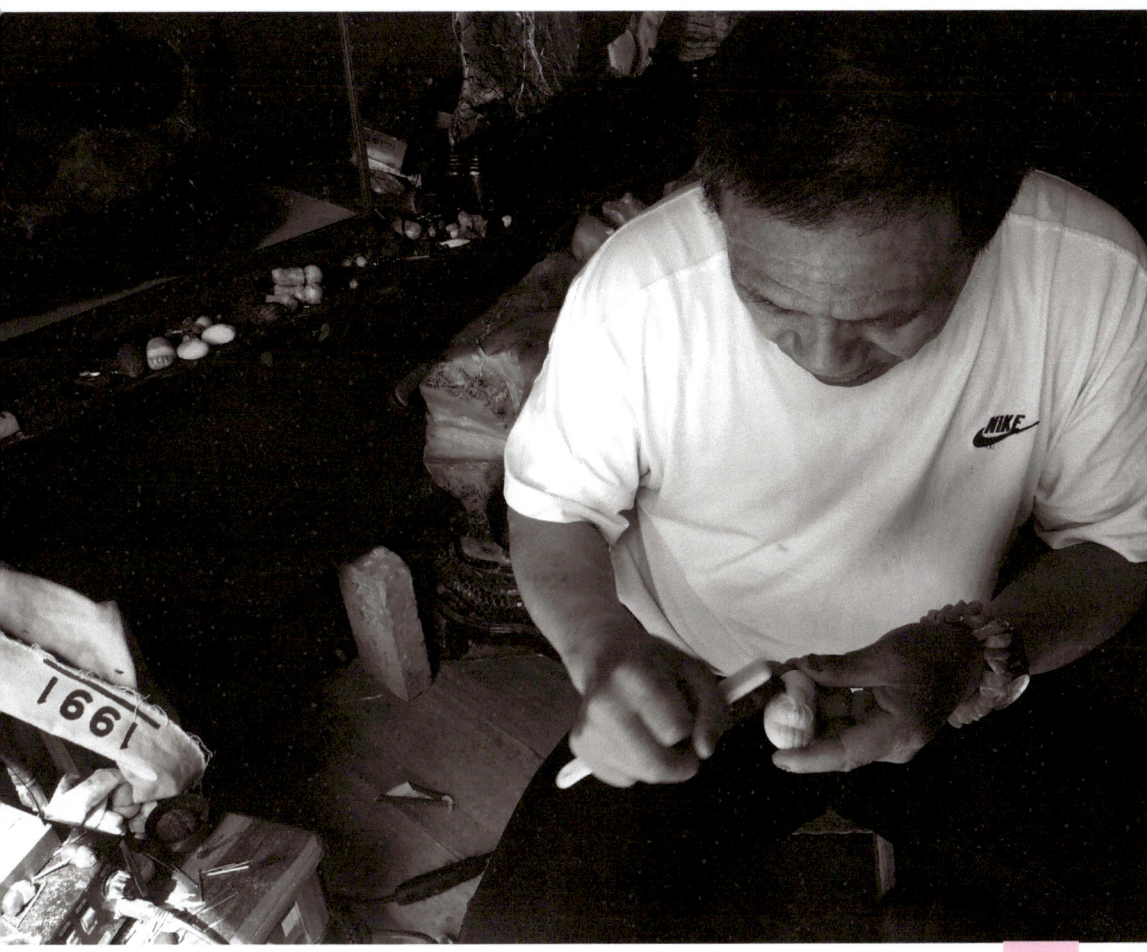

Figure 4.5
Meticulous attention to detail and surface texture is applied to a form being created by a craftsman in Wuxi, China. The already polished surface is continually brushed to ensure that the surface is perfect.

Smell

It may not seem important to consider smells, but if functioning correctly the sense of smell can be extremely sensitive and capable of triggering memories and recalling specific experiences.

Smells are useful in the search for ideas and in remembering past situations, but they can also help by enhancing or discarding an idea path through exactly the same mechanisms. As there may not be any physical pointers in recalling a smell, it is possible to receive mixed messages: messages that have the same trigger but that recall different events for different individuals.

Is it possible to be put off a product because of the way it smells? The dishonest used-car dealer may spray seats in a cheap car with classy scents to hide unpleasant odors, or perhaps to give an impression of a more impressive material, while the baker might ensure that the smell of freshly baked bread is directed at passers-by.

A consumer of a product may be influenced into making a purchase because of a smell that is emitted and the association that accompanies it. It would certainly be unfortunate if a product was to be rejected on the basis of a smell, but associations are powerful and should not be overlooked.

Materials that change their odor under different conditions will often result in different responses by the senses. Smells emanating from a material in a familiar setting may well change in unfamiliar settings and climates.

4.6

Figure 4.6
Experiences with similar odors can recall entirely different situations. A man repairing a fishing net in the street or a child in the park catching fish might have a similar odor associated to the activity, but can relate to very different experiences.

Sound

How important is sound? The sound of a door closing or a button being pressed can be an indicator of quality—or at least of a perceived quality. Sounds associated with a product are important and should also be carefully considered.

A switch that makes an audible click may be needed in some situations, but consideration should also be given to alternatives. What does the action of switching something on and off represent?

It may mean the start of a rest period for the user, or it may be an indication that something is about to occur. Whatever the reason, sound quality should appeal to the senses. The sound should not simply be a by-product of circumstance, but rather something that is integral to the user experience.

The integration of personal ringtones to cell phones enables the user to enhance their experience and to adopt their own personal preferences.

There are numerous products into which such integration could be incorporated, and whose experiences could be enriched with careful consideration to a note, pitch, tone, or melody.

Emotional responses

Happiness, sadness, calm, anger, fear, disgust: inanimate objects can instigate different emotional responses in different users through various inherent qualities.

A user's emotional response usually depends on the context of their encounter and intuitive associations. It is possible for an object to arouse, seduce, and stimulate beneficial feelings or emotions through appearance and memories. The intensity of these mental triggers can be further augmented or reduced through a physical relationship with an object, which may prompt emotional negativity.

The juxtaposition between mental and physical, positive and negative experimental emotions is difficult to predict, as everything and everyone is unique, but the manipulation of an object to assess positive and negative responses can identify and stimulate exciting ideas.

Adore

I like it, I want it, what is it? The reasoning behind why somebody likes something can be simple to comprehend, but it can also be complex and difficult to rationalize. The user may adore a trendy "must have" material, as it is believed that ownership improves their quality of life—a fundamental aim of design. However, a small grubby blanket or an apparently insignificant, tatty bracelet may also be adored by their owners if they are associated with childhood or past experiences. Such fond memories can also become more embellished as time passes. It is not possible to guess whether an object is adored by simply observing or thinking about it. A product needs to be placed in the correct context, and even then, what one person worships may be regarded as insignificant or peculiar to others.

Figures 4.7a and 4.7b
Designer Min Chen used the strength and simplicity of the Chinese character "gong" as a basic unit for multiple expression.

Added values

Unexpected

Chinese industrial designer Min Chen takes inspiration from Chinese calligraphy and the traditional Chinese arts and crafts, creating works such as Gong and Mu.

The simple form of the Chinese character "gong," composed of three strokes, is often a subcomponent of more complicated Chinese characters, and in the work of Min Chen this ideal is echoed in the constructions that are composed from the simple form.

Gong is a building block, a singular unit that can be manipulated and repeated to become multiple forms, each retaining the honesty and strength of the original Chinese character. The constructs are diverse, enabling infinite, unexpected forms to be created.

Figure 4.8
Screen created by Min Chen using the Gong form.

Added values

In the creation of the form Mu, designer Min Chen again took inspiration from Chinese characters and referenced multiple forms. "Mu," the Chinese word for "wood," represents a single tree as a single component, but as additional Mu components are added, Min Chen influences the meaning and the language changes. The placing of two Mu together represents the Chinese character "lin," meaning "forest," and when three are situated together they represent the Chinese character "sen," meaning "jungle."

The beauty of the form and the honesty of the material coupled with a comprehensive understanding of meaning and representation captures the imagination. The added value in the design is the intrinsic story of Mu, which, when it is revealed, manages to enhance the overall perception of the collection.

Figure 4.9a
Min Chen, Mu. The single component represents a single tree.

Figure 4.9b
The placing of two Mu side-by-side represents the forest.

Figure 4.9c
The placing of three Mu together represents the jungle.

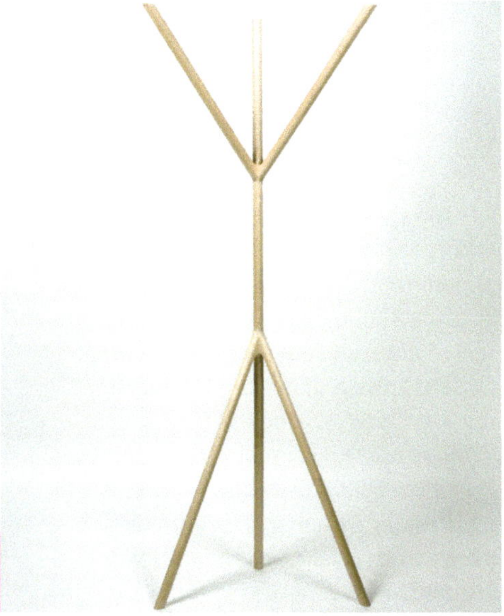

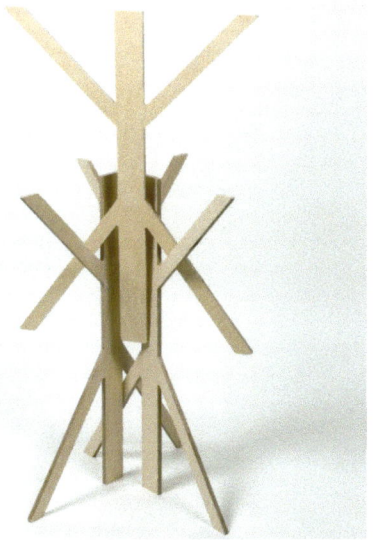

Added values

> "For the Candy collection we wanted to explore the colorful universe of our Sushi series and to express it through a glass collection."
> **Fernando Campana**

The Campana brothers, Fernando and Humberto, have constantly experimented with varied design approaches, exploring transformation and reinvention, while adhering to a faithful understanding of craftsmanship.

The Candy collection (Campana brothers, 2015) is inspired by their 2002 Sushi collection (including works such as Sushi IV Armchair), the similarities between a glass-making factory and a candy factory, and the colorful sweets in the markets of Brazil.

Hand-blown glass by Lasvit features in the Candy collection in the Lollipop table lamp and Sphere chandelier works. In a similar approach to the works Gong and Mu by Min Chen, with a faithful replication of a singular form, the Lollipop table lamp is the main influence in the larger works Sphere chandelier and Ring chandelier, while the individual craftsmanship associated with the Lollipop lamp ensures that each output remains unique.

Figures 4.10a and 4.10b
Candy collection: Lollipop table lamp (left) and Sphere chandelier (right), designed by the Campana brothers (2015) and produced by Czech glass-masters Lasvit.
Photo: Hana Klimova.

4.10a

4.10b

Added values

Taking inspiration from subjects as varied as a drain cover (Tattoo table, 1993) to sweets (Candy collection, 2015) the Campana brothers readily absorb everyday references from their immediate surroundings to steer and inform creative directions. The inspiration needs to be identified and then used creatively, but the Campana brothers have consistently demonstrated in their work that inspiration and influences can be found in the most unlikely of places, and can subsequently be transformed to create something incredible and sought after.

As with many of the Campana brothers' outcomes, the Tattoo table (1993), the Cobogó table collection (2009), and the tables of the Fitas collection (2013), they manage to demonstrate elements of added value when choreographed shadows appear under the works.

The manipulation and control of light to create the perception of a particular space being enhanced is a theme adopted in the subtle and graceful outputs Subconscious Effect of Daylight, Daylight Comes Sideways, and Surface Daylight 1 & 2, by Norwegian designer Daniel Rybakken. Exploring and appreciating the innate attributes of natural light and the multiple and varied lighting effects that are readily cast on an array of surfaces, Rybakken created lighting outcomes that capture the purity and innocence of daylight to effortlessly and calmly open up the spaces in which they are located or that they are able to influence. The works move away from conventional lighting approaches through the simple, everyday observation of natural light. Such lighting effects are common, where shards or mottled daylight openly dance on a surface. The inspiration is there for all to see, but Daniel Rybakken allowed it to influence his beautiful lighting series and further develop his thinking.

The inspiration seems so obvious, but it is very difficult for many to see, due to mental baggage associated with lighting curtailing or preventing the imagination from being triggered.

It is Daniel Rybakken's freedom of thought that enables him to create such imaginative and appealing designs. The work Subconscious Effect of Daylight uses a simple table to allow an unexpected, but seemingly familiar natural light to form on the surface below. The table is integral to the creation of the natural light, although the light with its distorted character appears to be distanced from the table and associated with a different and partially hidden source of daylight. In a different approach to the Campana brothers' Tattoo table, Rybakken incorporates the source of the light within the original object but manages to create the impression that it is natural light arriving from somewhere else.

Figure 4.11
Daniel Rybakken, Subconscious Effect of Daylight, 2008 (above).

Figure 4.12
Daniel Rybakken, Daylight Comes Sideways, 2007 (right).

Added values

The work Daylight Comes Sideways continues the experimentation with natural light and how it is perceived, introducing a fluctuating and subdued series of shadows on a surface. A feeling of movement and freedom is created that appears to be natural and instinctive. The space retains a simple feeling without being littered with elements that might be considered distracting or unnecessarily imposing. The lighting is no longer a conventional light, a static form, but a source of natural light that emanates into a space, altering conditions and feelings. The representation of daylight is portrayed as lighting in a state of flux, a condition that naturally alters the sensory perception of light.

The works Subconscious Effect of Daylight and Daylight Comes Sideways informed the Surface Daylight 1 & 2 collection by Daniel Rybakken. The works further explore the effects of daylight streaming into a space through the creation of panels that appear to have sunlight casting forms upon them. The impression created is the natural feeling of light entering the space, and it provides a sense of greater space beyond the immediate area.

The works differ from many approaches to conventional lighting in that they have an inherent awareness of natural light, and thus have an appeal that directly connects to the senses.

Figure 4.13
Daniel Rybakken, Surface Daylight, 2008.
Photo: Kalle Sanner.

Subtlety

To understand subtlety requires an understanding of judgment, an ability to appreciate subconscious detail, and a capacity to engage acute mental awareness. Subtlety in creative work is a delicate balance between acceptance and rejection; it is a characteristic between something being extraordinary and something being ordinary. The execution and craftsmanship associated with an object is fundamental to how an outcome is received.

Austrian sculptor Erwin Wurm has received international acclaim for works such as Fat House (2003), Truck (2005), and Fat Car Convertible (2005). As a sculptor, Erwin Wurm has an ability to intimately understand objects, a developed understanding of meaning, and a sophisticated sense of balance.

Figure 4.14
Erwin Wurm, Fat Car Convertible (Porsche), 2005 (edition of 3).
Photo: Erwin Wurm and Xavier Hufkens, Brussels.

Control

Fat Car Convertible (2005) exhibited at the Xavier Hufkens Gallery appears to be anything but subtle, with its oversized body that seems to be a direct contradiction to the perceived view of conventional convertibles, but the sculpture is undoubtedly subtle. The work is detailed in a manner that might not be openly apparent, but the control of the work is explicit.

The subtlety within Wurm's Fat Cars is distinctive and presents a language opposite to contemporary thinking, but it is also a language that carefully exudes a set of considered alternatives, such as "satisfied" and "serene," in addition to being seemingly carefree and relaxed.

Such characteristics are attained through a confident, subtle appreciation of form, despite any immediate, literal impression that the Fat Cars are anything but subtle.

The temptation to do more and to continue to tinker with an object beyond reason can create confusion and detract from the integrity of an original thought. South Korean designer Miyoung Nam understands control and continually manages to create visually uncomplicated outcomes such as Snow (2016) and Shard (2016). The lighting designs appear simple and elegant but have been carefully crafted and managed to remove all unnecessary markings, including any visible evidence of how the components are inserted into the forms. There are no external markings of any kind. The core components that contribute to the arrangement of Snow and Shard are of primary importance and the subtlety in their arrangements and execution ensures a considered outcome is achieved. Control in the design process is central to success and any temptation to unnecessarily overwork an idea should normally be avoided.

Less is more.

Control

Control is also evident in the work Pole (2015) of Miyoung Nam, where all components and practices have been kept to an absolute minimum but where daring, experimental thinking is still able to coexist comfortably with simplicity.

An adjustment to preconceived and hackneyed proportions introduces an unforeseen language that is able to stimulate both curiosity and intrigue. The confidence to challenge and defy what might inappropriately be considered to be a correct arrangement, simply because something has always been done in a particular way, introduces an opportunity that can prompt alternative directions. Considerable care has been taken to ensure that the work is uncomplicated in every aspect, including the reclaimed components that remain in their original state when assembling the towering structure. There is no need to deviate from an idea and add more if it is not needed. The Pole (2015) light makes a statement through its overall form and any amount of additional work to individual components is always likely to be secondary to the overall structure.

Miyoung Nam arranges clusters of towering poles to create imposing structures that illuminate space in an unexpected and untraditional manner.

The work is essentially a component, that when it is replicated, has the potential to allure and excite. If the work were overly complicated at the outset, the essence of the idea would be lost and it would be difficult to assemble multiples of the design without it appearing too busy. Controlling the building blocks of design is necessary if outcomes are also to demonstrate an element of control.

Figure 4.15
Miyoung Nam, Pole (2015) light.

Control

Challenging anticipated perceptions can be achieved in broad terms through a simple and unexpected modification of just a single component. Such adjustments have the capacity to transform a relatively insignificant object into a captivating and intriguing one. Such simple changes can make a dramatic impression, and in doing so reinvent original narratives that can elevate the mundane into a completely different arena.

The introduction of multiple legs in Walker by Oliver Schick Design (2007), the large soft seating component in Barbapapa (2006) in the narrative work 100 Chairs in 100 Days by influential designer Martino Gamper, and the mixture of unexpected materials and form in the musical-themed Showtime Armchair BD (2007) by the Hayon Studio for BD Ediciones de Diseño all demonstrate an ability to challenge perceptions and introduce alternative directions through subtle and considered change.

The lighting designs Bamboo Blue XL (2014) and Sugar XL (2014) are inspired by the observations of bamboo and sugar cane. The lights are constructed from the multiple use of a single unit that has not been purposefully produced. The components are abandoned thread spools from the textile industry. The individual spools are simply pushed inside each other to create tall structures that take on a form that is representative of bamboo and sugar cane. The development of these particular light structures does not require any modification to the individual components and no tools are needed in the construction. The design has been carefully composed to capture the characteristics of bamboo or sugar cane.

In Sugar XL (2014), where the separate components are connected together, they create a double thickness of the material and when illuminated from within the connections can be easily seen appearing to be similar to growth rings or the incremental sections in sugar. The outcome is not incidental but has been carefully considered. If the sections were not constructed through push-fit and used a different method of construction it is likely that any additional component would be visible when the light is illuminated.

Figure 4.16
Y&Bramston Sugar XL (2014) and Bamboo Blue XL (2014) inspired by sugar cane and bamboo.

Respect

The degradation of an adored product can lead to subconscious neglect and a distancing between the user and the object. Up until a product receives its initial tarnish, perhaps a scratch or dent, it is usually respected and cared for; however, when something does occur, the object can be relegated to a league of lesser importance, and user focus can become tarnished and directed at a possible replacement or an alternative item. A caring attitude can begin to dwindle with a blemished product, and the love and respect it may once have been shown can be forgotten. It is, however, not uncommon for products of a decommissioned status to actually become more useable in practical terms.

After the initial shock of damage is acknowledged and alleviated, an object can actually begin a sensible existence: an existence in an environment where accidents do occur and are recognized. The scratch or bump usually only impacts on an aesthetic level rather than on operational capabilities, and if such battle scars are actually understood they can add a unique, distinctive character.

The idea of a design being visually unattractive due to knocks and bumps can be contrasted with an object's sudden practical status. The concept of scratches, bumps, rips, and dents can often be seen as a desired feature and inspiration for originality; as well as an opportunity to personalize.

With an individual aesthetic and ideas about extended existence, Mehmet Erkök used a Nokia 3210 in experimental work, replacing and altering original components to develop a unique, personalized product.

The cellphones of Mehmet Erkök explore rechargeable cells, operational qualities, and usability through the use of visible batteries, an ability to "dress" the phone for specific requirements, and an acknowldgment of what Erkök describes as "stereotypical learning." The phones have distinctive personalities that demand attention and respect.

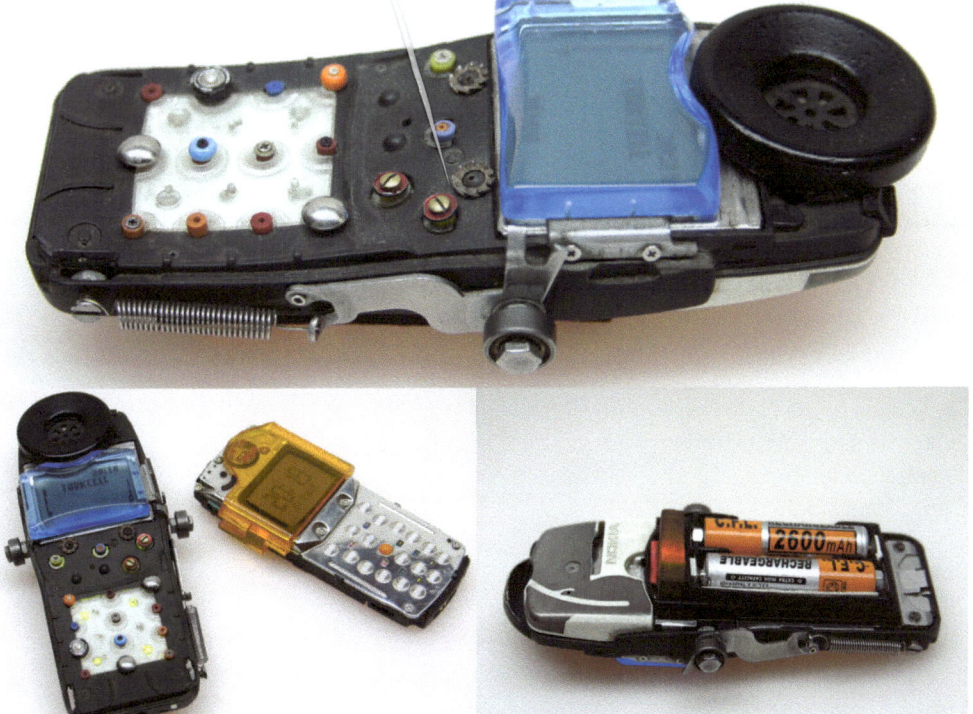

Figure 4.17
Mehmet Erkök, "Custom" or "Hacked" cellphone.

132

Adopting Responsibility

Designers have to push the imagination of others, be daring and make the most of exciting potential markets; to present opportunities by doing what others have not done. But they also have a duty of care. They have a responsibility to ensure that their outputs are acceptable at many levels and in multiple contexts. There needs to be an appreciation of the resources available and a consideration of the broader implications that a particular idea might have. There is a moral, ethical, and social responsibility associated with design, where all directions need to be continually evaluated.

Sustainability

Designers have a responsibility to improve life, to make it more enjoyable, more efficient, and more entertaining, but they also have a responsibility to achieve this without introducing any detrimental characteristics in their products. The creative stages do not need to be compromised, but they do need to be carefully considered. Consideration needs to be given to all the stages in the development process on how something could be improved and how any negatives, whether immediate or long-term, might be avoided. The introduction of such a constraint in the creative process should be considered as an indicator of possible opportunity rather than a hindrance. The designer and the creative industries have a moral, social, and ethical responsibility that needs to be continually acknowledged and accepted.

An appreciation of responsibility ultimately complements a design and its broad acceptance since there is usually a beneficial story to tell. Identifying an exciting idea and being enthusiastic about it is undoubtedly a good thing, but it is also possible to become so enthralled by the momentum of an idea that the fundamental and broader implications become overlooked or ignored.

The design process is a problem-solving process and the problems that need to be solved are wide and varied. A successful design should include an awareness of key factors, such as resource usage, and the designer should be able to steer an acceptable path in terms of sustainability.

5.1

Figure 5.1
Shao Luya, Zhang Yahan, Ma Lijuan, Pan Qiuguo, Huang Shuyin, Zhao Ping, Li Yuxin, Dong Wei, and Sun Jing, Yoghurt Pot Chandelier, 2014. Photo: Qicong Lin.

> "We are all astronauts."
> **Buckminster Fuller, *Operating Manual for Spaceship Earth***

Resources

In 1968 American architect, innovator, author, and creator Buckminster Fuller stated that "We are all astronauts." The statement suggested that the global population was onboard the Planet Earth Spaceship, and that it was unable to stop for additional resources, indicating that as a community, the "astronauts" need to appreciate the value of resources because when they are gone they are gone.

The essence of the statement remains relevant in the twenty-first century and global resources are something that need to be considered carefully in the generation of ideas.

The Yoghurt Pot Chandelier, designed by the Tsinghua International Foundation program in Beijing, recognizes the beauty of the glass yoghurt pots that are so readily available on the streets of China. Although the glass is not of a particularly high quality compared to that made by specialist glassmakers, the designers managed to create a series of experimental arrangements that promote the whiteness of the vessels when in close proximity.

In using simple yoghurt pots, the arrangements collectively hold many stories. The story of the street, who used the yoghurt pots, where were they going and so many other questions create a hidden value that would be absent if virgin material had been used. Although the components of the design are familiar to people accustomed to life in China, the form that has been used is unfamiliar to other cultures, and so there is no preconceived understanding.

The use of found objects or rescued materials is becoming increasingly important in the design process. Materials that have been rejected for minor reasons, such as color imperfections or because of a slight manufacturing fault, can offer great potential and can be turned into outcomes with meaning and an inherent sense of responsibility.

Sustainability > Tinkering

A preformed, previously used material should not present a problem in the design process if it is considered with the same imagination with which any other material is considered, but it also has the potential to offer so much more than a virgin material.

Stuart Walker, Co-Director of the ImaginationLancaster Design Research Center, is Professor of Design for Sustainability and has produced various reuse works such as Lather lamp and Wire light. Such works use simple, understandable, pre-used materials, to create meaningful designs that are desirable on multiple levels and which are capable of being readily maintained.

Although such works appear simple, the thought processes behind them are carefully considered. These works are not simply an amalgamation of various components randomly gathered, but rather a comprehensive understanding of need, an appreciation of sustainable practices, and a demonstration of possibility that recognizes the importance of valued resources. The Lather lamp design uses a discarded soap bottle as a predetermined form, a steel rod, and cast concrete. These materials are simple, they have honesty and transparency about them; there is no hidden agenda, and because of this they have an elegance and appeal that many more complicated and irresponsible solutions, blinkered to core issues, are unable to capture. Wire light takes the concept of simplicity further, but in doing so the design is not lessened or compromised: it becomes an object of functional beauty.

There is no complication to the outcome; it simply puts a light source where it needs to be effectively and efficiently. The ability to really understand need, to strip back an idea, to remove resource-intensive processes and to make a design understandable to all is undoubtedly at the core of good design thinking.

It is surprising that such effective, efficient, and simple works, creations that can tell many stories and are so wonderfully elegant are not a catalyst for substantially more works. There is too often a desire to source original material, to overwork a design and to create unnecessary outcomes when simplistic and carefully considered solutions would be more appropriate.

"In keeping with the principles and practices of design for sustainability, Lather lamp and Wire light are attempts to demonstrate that elegance and utility can be achieved by minimal means at the local level, and in ways that lend themselves to visual clarity and functional comprehension. Such comprehension facilitates product maintenance and repair. In turn, maintenance and repair allow for product longevity—and product longevity combined with localization contributes to the development of a cultural, meaningful material world while simultaneously reducing waste."
Stuart Walker, Professor of Design for Sustainability; Co-Director, ImaginationLancaster Design Research Center

Figure 5.2
Stuart Walker, Lather lamp. Reused soap bottle, hand-formed mild steel bar, concrete casting, general-purpose electrical parts.

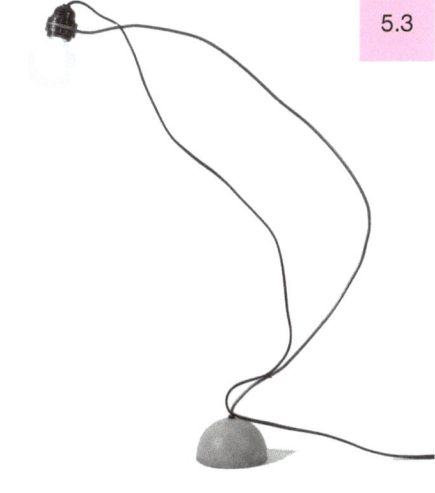

Figure 5.3
Stuart Walker, Wire light. Hand-formed mild steel bar, concrete casting, general-purpose electrical parts.

Openness

Brazilian designer and craftsman Paulo Goldstein manages to combine a multiplicity of creative, divergent experiences and an appreciation of craftsmanship in collections, such as Repair is Beautiful (2012), where control is on a personal level with the individual. Broken objects that impede or deter functionality and promote frustration are ingeniously renovated with an approach that introduces intrigue, authority, and meaning. The works of Paulo Goldstein in the Repair is Beautiful collection are not simply repaired to their original state, but repaired in a manner that redirects frustrations associated with non-functioning, broken objects, and challenges the original values of the object.

The Repaired Anglepoise 1 work had to respond to a series of problems, which impacted on further modifications and adjustments. The work ultimately becomes individual, personal, and meaningful.

Figure 5.4
Paulo Goldstein, Repaired Anglepoise 1. Ingeniously created to avert the frustrations created by a non-functioning, broken object.

The Repaired Director's Chair by Paulo Goldstein has a unique language and identity, but is restored to its original functionality, having been sourced as a broken frame with a missing backrest, and the scars of the various inadequate repairs of a previous owner. The Repaired Director's Chair is a statement of achievement, an object that has not been abandoned, and despite appearing to be in traction with its tension cables and various rigging assemblies, the designer has been able to overcome the innumerable challenges presented through sensitivity and craftsmanship.

It is too easy to discard an object, to abandon it without thought and to ignore the consequences of such actions. Rescuing broken works informs creative directions and presents a series of challenging and new constraints that can be overcome through simple ingenuity. The structural problems that Paulo Goldstein resolves through ingenious solutions are akin to the adlib structures created by Matthias Pliessnig (see Chapter 1) when simple found items are brought together to inspire and challenge.

5.5

Figure 5.5
Paulo Goldstein, The Repaired Director's Chair.

Sustainability > Tinkering > Inspirational

Honesty

The generation of second-life products is a burgeoning area of interest, where items are not only designed for their initial or primary use, but are also considered for a potential secondary life that does not require materials to be recycled.

In designing an original object, consideration should not only be given to the processes associated with how the idea will emerge, but also to what will happen to the object when it is no longer functional. Designers such as Paulo Goldstein manage to engage creativity and craftsmanship in developing new identities for objects where there has been no original thinking directed at what the object would become after it ceases to function correctly.

Thought can and should be given at the initial creation stage as to how individual components of an emerging product can be responsibly disposed of, or how they can seamlessly morph into another product without any exhaustive processes.

Figure 5.6
Zhang Yuxi, Found light, 2014.

Tangshan-based designer Zhang Yuxi created the lighting design Found (2014) from locally sourced and discarded materials. Combining imagination and creativity with responsible design, Found is comprised of opaque waste plastic, an abandoned transparent glass vessel and a vinyl toy duck. The randomly selected ingredients for the design were constructed so that the light source above the vessel would adequately illuminate anything contained within. The object within the vessel could in reality be anything that is personal to the user, and might include a toy, a personal letter, a photograph, or something else with an individual meaning. The assembly of Found is not intended to be complicated and is honest about the use of simple fastenings to connect the entire work.

Found is a study that aims to create acceptable lighting through responsible practice. The rawness of the light becomes the light's main attribute, since it allows it to be original, maintainable and personal, in a similar manner to the inspirational approach of Mehmet Erkök (see Chapter 4).

Yellow Soldier light (2014) is also constructed from abandoned materials and in particular recognizes the advantages of cheap plastics. Observing light passing through the colored lids of soda bottles initially influenced the design. The lids of the bottles acted as simple color filters allowing color to be added. The action inspired the beautifully simple Yellow Soldier light, which uses cheap, translucent colored plastic items, situated under a prominent light source to poach color away. The light allows for unique personalization.

Figure 5.7
Zhang Yuxi, Yellow Soldier light, 2014. The original version of the light, following investigations with colored soda bottle lids, used a cheap, yellow plastic toy soldier as a color filter.

The design works of Zhang Yuxi were created from found objects rather than objects designed for a second life when their initial life is extinguished. Although second-life products are perhaps not a solution to a global problem associated with excessive waste material, they are worth considering in the idea-generation process, and they have many possibilities. Many design products are too readily abandoned after their immediate use, but so many products have such similar forms and attributes that evolving a product into a secondary product should be considered more often.

It is not too problematic to design a plastic container for detergent in a way that makes it a viable lamp in a second life, or for a plastic container to become a stool. This is designing backwards, where the secondary object is considered as the primary objective and the initial object simply becomes a carrier for the successive idea.

Considering a design beyond the immediate confines of a single lifetime introduces the possibility for mixed languages to be explored and for multiple options to emerge. Hybrid products, which share core characteristics but which are also fundamentally different, are able to emerge.

The innocuous plastic detergent container might adopt a particular surface texture that enables it to be transformed into a viable lighting solution. The readily available plastic oil container that appreciates the potential of a second life can be slightly modified to become a garden planter or watering can. The opportunities for taking an original idea and allowing it to evolve into an additional idea are compelling. A plastic oil container could be created as an object of beauty if it was to evolve into something with a longer life expectancy.

The potential of materials is consistently recognized and demonstrated in the creative works of Italy-based Vibrazioni Art Design. Predominantly sourcing abandoned steel, and in particular screen-printed metals from the petroleum industries, Vibrazioni Art Design create alternative contexts through repurposing, understanding, and exploiting the beauty of the materials.

Surface imperfections and material vulnerability, in addition to a variety of colorful screen-printed sections, present a creative opportunity that is supported through the honesty of fabrication. The scars of construction are openly visible, rather than shrouded or hidden away, and contribute positively to the industrial narrative.

Blemishes and defects instil curiosity and have meaning. Defects can be converted into accepted and actively sought after characteristics when a creative ethos is repositioned. Recognizing the potential beauty and realistic practicality of used objects to inform a creative direction requires a creative perception and a desire to succeed. Abandoned materials have much to offer but are too readily dismissed.

Figure 5.8
Alberto Dassasso, Vibrazioni Art Design, Avio Chair. Vibrazioni Art Design outputs are shaped and meticulously worked into their individual forms that capture and promote the beauty of the surface print.

Figure 5.9
Vibrazioni Art Design, Honda Trequarti motorbike. Salvaged materials are worked to generate an unexpected industrial aesthetic that exploits the direct relevance of the screen-printed metal featuring the logo of the automotive lubricant manufacturer Pennzoil.
Photo: Callo Albanese.

The Freitag brothers, Markus and Daniel Freitag, created the original FREITAG messenger bag, the F13 Top Cat, through personal experience in 1993.

As young, observant designers, cycling with their artworks in Zurich, they ideally needed a strong, durable bag that would be water-repellent and robust but that would also appeal to them as individuals.

Observing trucks on the Zurich highway they realized that strong, weatherproof truck tarpaulins would make the perfect material for the construction of such a bag, since the requirements for the bags were fundamentally similar to those of tarpaulins.

Understanding problems from personal experience is a significant advantage in the generation of an idea, but it still requires an open mind and an ability to make the necessary connections.

The connection that the Freitag brothers were able to make between truck tarpaulins and a potential messenger bag solution is obviously significant, but a viable outcome still needed to be created. The creation of the original FREITAG bag using a truck tarpaulin, bicycle inner tubes and seat belts resulted in a bag that not only had the properties required to solve the immediate problem of protecting artwork, but since the materials were associated with transit and haulage the industrial aesthetic of the bag was ideal. Despite emerging from and then entering a second robust, courier life the craftsmanship and details of the FREITAG bags remain paramount, and details already featured on the tarpaulins, such as pressure-bonded fabric, is naturally captured in the bags.

Since the tarpaulins are all different and sourced from diverse areas of Europe, bags that are handmade become unique items with individual color and detail combinations. As the design of the tarpaulins changes, the bags naturally follow and in doing so remain fashionable.

Figure 5.10
A FREITAG F13 Top Cat messenger bag, following the basic design and principles of the original bag, which the brothers created at home on their mother's sewing machine using tarpaulin washed in the bath, bicycle inner tubes, and car seat belts.
Photo: Peter Würmli.

Simple

Observing the removal of tangled and damaged polypropylene packaging straps used in the haulage industry, designers Lucia Lopez Garci-Crespo and Zhou Hui used the properties of the material to create a series of Fishingsticks, suitable for children to use in local pools in China. The flexible, waterproof, and strong characteristic of the packaging straps were manipulated through experimentation and play to create a braided Fishingstick using several pieces of the resistant and purposefully defiant material. A simple colored line attached to the end of the Fishingstick accompanied by a delicate feather and a small knot at the opposite end of the line for attaching bread provided a Fishingstick that would feed fish and allow the child to encourage the fish to chase food without harming the fish.

    The design is simple, yet effective, and recognizing that the properties of an unwanted material can become the basis for a sought after and uncomplicated outcome is a necessary trait in idea generation.

Figure 5.11
Lucia Lopez Garci-Crespo and Zho Hui, Fishingstick.

Inspirational

> "Since 1990 I have been occupied creating new life-forms."
> **Theo Jansen**

Resources

There are numerous situations where resources are finite but solutions to immediate problems need to be sought with basic materials. Such practice is abundant and can be evidenced everywhere. Individuals who generate solutions through need are true inspirational creators, since they express outcomes through their own inherent understanding and personal experience of what is necessary.

Creator Theo Jansen adopts the approach of creating from everyday materials to construct Strandbeest: beach animals that are composed of old soda bottles, and yellow plastic tubing normally used for housing electric cables. Initially just playing with such accessible materials and experimenting, Jansen initiated the ambitious process of developing herds of kinetic beasts that are able to continually roam on the broadly flat terrains of the beach.

Transforming the inanimate into potentially animate forms, capable of living, Theo Jansen created the original and unfamiliar Strandbeest life forms with familiar animalistic traits, such as feathers, which capture the wind to enable the movement of the animals' abundant feet and legs.

The Strandbeest have matured since the initial outset, having adopted an ability to use their impressive and forceful wings to accumulate wind in soda bottles for use when there is insufficient wind available for their beach migration, away from the surf towards more protected regions.

Figure 5.12
The dynamic Strandbeest, by creator Theo Jansen, continually roam the beach in herds, walking on the wind and sensing dangerous surf.
Photo: Theo Jansen.

Figure 5.13
Theo Jansen, Strandbeest.
Photo: Theo Jansen.

150

Evolving the Reality

Blue-sky thinking is an important aspect of the idea-generation process, but any idea is simply an indicator of possibility that needs to be nurtured and developed if it is to succeed. An idea is ultimately a thought with potential that can arrive suddenly due to inherent knowledge or it can be carefully identified through doing various activities.

Blue sky

Reality

An idea is fundamentally a seed for potential. It is a thought that needs to be nurtured and supported if it is to develop into something that is ultimately desired. The ability to capture an idea or to identify a particular thought can be encouraged through many varied activities and experiences, but where there is creative understanding and instinctive knowledge, an idea can simply arrive without being consciously encouraged.

An idea that suddenly appears to an individual has the potential to be as influential and significant as an idea that has had to be teased out.

There are many formulas for the design process, but the creative pathway is not as straightforward as some might suggest, and a designer needs space to maneuver thoughts and explore unexpected directions and encounters. An intuitive creative, who is naturally able to embrace and respond to multiple encounters and adapt their pathway accordingly, is a creative, who is more likely to succeed than the individual following a rigid route without any flexibility.

Designers require information, either inherent or researched, to activate the mind and to consider emerging opportunities. Exploring the "if only" and "what if" scenarios associated with **blue-sky thinking** is of fundamental importance, since these are the questions that are capable of moving boundaries and revealing beneficial directions to follow. Identifying an exciting idea and being enthusiastic about it is a good thing, but it is also possible to become blinkered and so enthralled by the momentum that basic elements are overlooked. Is the design actually that prudent?

Some of the basic questions that should be considered when thinking about an idea are:

- Is the function realistic?
- Is it practically realistic?
- Is it logistically realistic?
- Will it be realistic morally?
- Will it be realistic ethically?
- Would it be realistic to create?
- Would it be realistic to market?

A rethink regarding a particular problem area does not necessarily mean a complete redesign, and minor alterations can have significant benefits without being too detrimental to the desired outcome. The ideas stage is a continuous period of evaluation in which a seed of potential is encouraged to develop.

The Tune 'n Radio proposal was created by Wouter Geense Design Studio in the Netherlands. The design recognizes the importance of personalization, with a radio design that is unique and customized by the user. The radio invites the user to engage in the design process and direct the outcome.

Figure 6.1
Wouter Geense Design Studio, Tune 'n Radio, 2005.

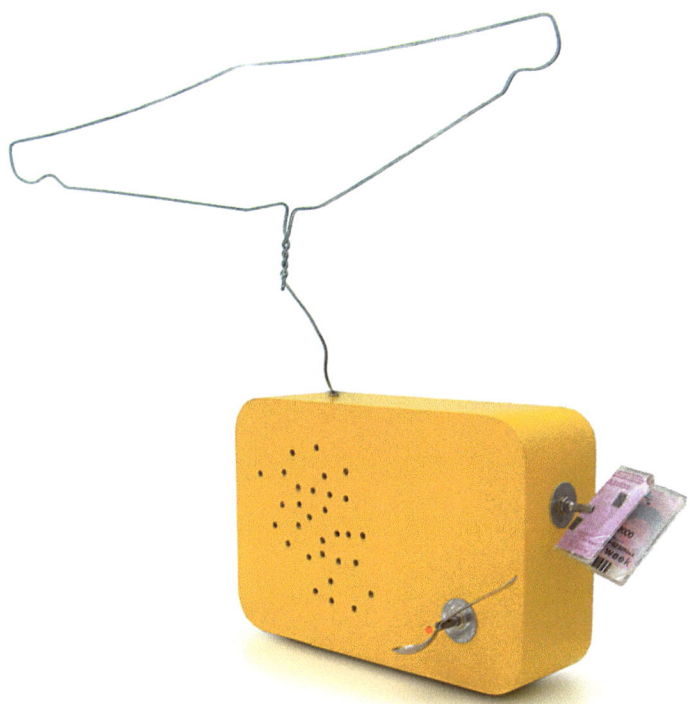

Development

An idea is a starting point, a point that may be the initiation of a proposal, but first it needs to be interrogated and scrutinized.

Identifying an idea and then developing it is by no means the end of the story. A product needs to be much more than a physical commodity. A product should engage the user, appealing emotionally and intimately, to ensure that interaction is positive. A product needs to connect in a manner that is acceptable and personal; the consumer wants more than a tangible thing—he or she requires exposure to an experience, an experience that aligns with his or her own beliefs, standards, and aspirations. It is no longer very difficult to find products that are comparable in terms of physical beauty or function; the distinguishing difference is providing a beneficial experience and making a unique impression.

The experience should not be considered after other development matters have already been decided on, but rather at the outset—it is an intrinsic and fundamental issue. A brand experience should be of utmost importance at the initiation of an idea.

Opportunities to fascinate and captivate an audience on an individual level are significant and are more aligned to nurture than manufacture.

Every single aspect of a product needs to be carefully evaluated and understood. The experience of a brand is influenced through physical, mental, and sensory encounters. If an experience is perceived as being contradictory, incoherent, or irregular then a unified message is not being communicated effectively and the response can be damaging. Everything needs to be considered carefully and thoroughly to formulate an impressive dialog that has meaning and consistency. Consideration and appreciation of the brand experience should be understood from the outset to influence idea direction, formulate desirable objectives, and recognize important constraints or themes.

It is necessary to have empathy with a brand and to appreciate what it stands for and why, as much as it is to have an understanding of a target audience and really comprehend what they want.

Although a brand experience must work on a personal level, it should also perform well on all other levels on which it will be encountered. An evolved reputation of a brand, positive or negative, will almost certainly impact on the individual consumer; attention must be focused on everything associated with the brand if understanding is to be enhanced, as a ready-formed incorrect impression can be difficult to overturn.

Packaging

The packaging of a product is a subject that needs to be considered with ever-increasing care, attention, and understanding.

Packaging is an integral part of the design process and should not be an afterthought. It should be subjected to as much rigorous investigation as the item it surrounds, protects, and promotes.

Why package something that is already packaged? A strange question? Packaging does not simply mean plastic bags, card, and polystyrene; it is anything that surrounds something else, a skin securing and protecting more vulnerable items.

It is not unusual for packaging to be a physical component of a product; something that is not discarded but rather embraced can add value. Natural materials are often preferential to synthetic materials, although creative thinking will reveal exciting opportunities that are not harmful to the environment and can actually enhance an idea.

The need to consider packaging carefully has become an influential design constraint in the generation of ideas. Innovative packaging has to be explored, as audiences are starting to appreciate efforts to reduce the environmentally detrimental effects caused by some packaging.

The notion of reuse is no longer merely a desire, but a necessity; a necessity that is creating a wealth of ideas. Inspiration for packaging-related ideas is everywhere, particularly in regions where materials are limited and where there is a greater number of design constraints.

What does it say about an individual or a company if there is no consideration to environmental impact when designing or purchasing a product? Target audiences are becoming increasingly aware of social responsibilities and associations.

1 Illuminated beauty

Understanding the character of a material through experimentation and play can provide insight for potential directions and outcomes. The diffusion of light through cheap, translucent plastic products such as a bowl, a plate, a toy, or a bottle cap is a characteristic that can be exploited in the creation of various lighting solutions. Abandoned colored plastic is simple to source, and in situations where such items are weathered or distressed the opportunity for experimentation with light is enhanced. A scratch or a series of marks on the plastic does not need to be seen as detrimental, but rather can give an understanding of the object's former purpose, and might even suggest a particular direction to adopt.

Collecting a range of cheap and readily available colored plastic products, conduct simple tests to observe light passing through the material and the impact of the light on its immediate surroundings.

The creation of a lighting design is fundamentally about the control of light and what the designer is able to make the light achieve. Observing colored plastics in different arrangements should suggest a potential pathway for developing a translucent plastic lighting solution. The repeated use of identical objects or items with similar characteristics can create an outcome with greater impact than a single item illuminated on its own. Allowing an idea to evolve through experimentation and play creates a beneficial foundation for developing a wonderful lighting solution.

2 Hidden seating

Street traders often adopt abandoned and frequently broken pieces of furniture, and subsequently embark on a journey of repairing, modifying, and personalizing the items to achieve basic comfort and durability. The condition of the furniture and the manner in which it is both abused and cared for are valuable indicators for generating potential ideas. Seating is often wrapped in offcuts of foam or tape where there is greatest contact with the user. Legs to the seating might be eclectic and repaired, but they are also one-off solutions that might accommodate a particular terrain or environment.

The appearance and thoughts that are expressed are often intriguing and curious and it is difficult not to be fascinated by the array of innovative repairs that have been made.

Due to the hybrid characteristic of these chairs, specific features might be removed or returned to their original place at the end of trading. A box, some blocks, and perhaps a piece of packaging foam might be the essential components of a particular seating solution, but when the item is deconstructed and items are returned, the chair effectively disappears.

Exploring your immediate surroundings, create a range of seating solutions that can be constructed from random items that are not naturally associated with seating. The construction should not involve complicated processes, and the completed works must be capable of being easily deconstructed and returned to the surroundings unnoticed when not in use. To prompt design ideas, "see" what individuals sit on when they do not have a traditional seat to rest on.

3 Unexpected elegance

Inspiration does not need to be sourced in obvious locations, and it is often the case that unexpected references provide the most interesting directions to consider. Designing a particular object with reference only to that same genre of objects prevents the imagination from being stretched and simply suggests directions that have already been explored.

Looking to seemingly unrelated areas for inspiration and appreciating that it is not necessarily the whole of something that is being observed, but a particular component or asset that is important, throws up multiple thoughts.

Interdisciplinary or cross-disciplinary references can provide valuable triggers for thinking. Eclectic references from a variety of sources can, if managed appropriately, provide some of the most interesting outcomes for steering an idea.

A visual review of some of the leading fashion journals where creative photography captures a particular mood can kick-start the imagination for a design that might appear to be completely unrelated to fashion. Seeing beyond a photograph on the pages of a fashion journal and deconstructing the essence of the outputs presents an array of potential directions. Fashion images can suggest delicacy, structure, layers, elegance, exaggeration, flow, excess, and many other ideas that can be beneficial to design thinking.

Selecting a single fashion image, aim to identify the main characteristics of the photograph and look to transfer some of these features into a design for a domestic piece of furniture. The outcome should be feasible and original, and should communicate an unexpected elegance.

4 The 99p orchestra

Designers need to be constantly aware of their surroundings. They need to be aware of what is actually taking place and be able to notice the potential in things that others might not see. Design is not an "on and off" activity, but rather a constant process of seeing and understanding. It is a process of thinking "what if" and "maybe."

The continual collecting and storing of artifacts for reference is of fundamental importance, as interesting items, or parts of them, can feed the imagination.

An info dump is the presentation of many varied items of interest to others for evaluation and consideration. The only common thread between the items presented might simply be that they were intriguing and curious.

Many objects sourced for an info dump can be free or inexpensive. The collection might include novelty items—considered kitsch by others—or they might be objects sourced in a charity shop, simply for a particular attribute or a sound that they might make.

Collect as many items as possible that create an interesting sound and that are not immediately associated with instruments found in a traditional orchestra. Imagination is needed to identify such sound-making artifacts, but it is also necessary to engage the imagination to think of where potential items might be found. The individual items that are collected should not cost more than 99p each, and where possible should be obtained for less. The collection is about creative thinking, not physical value.

In addition to the sounds that are possible with the various items acquired for your 99p orchestra, explore other characteristics such as structure, tone, harmony, and emotion, and also look to represent such traits through your found items.

Evaluate the items collected and identify any areas of particular interest that might inspire a musical instrument.

5 Harmony and tension

A freedom to explore through play and express thoughts through the creation of simple structures broadens the imagination. The absence of a specific agenda and immunity to pressure allows for thoughts to flow naturally and to be maneuvered comfortably between emerging options.

Observing elements such as stress, balance, pressure, harmony, tension and compression in wild vegetation, look to recreate these characteristics in a series of simple maquettes. The studies should be constructed from an amalgamation of inanimate objects that can be sourced within the immediate working area and should not take any longer than twenty minutes to create.

6 My place

Inspiration is found everywhere and areas of interest should be noted. Being creative in a work area where there is an absence of visual inspiration is an unnecessary and onerous challenge. Visual references do not need to be literal, since it is often lateral or abstract traits rather than creatively constrained references that offer more realistic potential.

A sterile workspace with no obvious visual references can hinder creativity, but it is also important not to overload an area with a single train of thought, since this can also inhibit the ability to attract diverse thoughts.

At the outset of a design journey, multiple and varied references that are able to portray a broad array of messages are likely to be the most compelling and convincing.

Concentrate on creating a viable and creative workspace with multiple visual references that are of interest for a variety of reasons. There should be a story or a reasonable argument associated with all the images that are selected for purpose. Aim to continually update the images as the design process unfolds, and maintain the strong undercurrent of interest. The workspace references should continually excite and feed the imagination.

7 Hello

A designer needs to be proactive and able to undertake activities that might take them beyond the immediate comfort zone of the studio. It is necessary to meet a broad array of individuals and to engage in frequent conversation without demonstrating bias or preconceived views, in order to gain understanding from an alternative or comparable standpoint. Instigating a conversation can be made easier with the use of an unrelated artifact that can remove any barriers of concern. Chinese designer Guan Ziyin (CAFA IFC 2014) used an old street worker's stool to engage different characters she met, by asking them to simply sit on the stool and have their photograph taken. The practice is simple and is able to facilitate valued dialog with individuals of interest.

A stool provides the ideal opportunity for engaging in dialog, but many other carefully considered items can be used equally well to stimulate a conversation and break down a preconceived barrier in order to try and ascertain a particular view on something.

In a group, collectively discuss and select an unusual handheld curiosity with the intention of the artifact facilitating a particular conversation. The item selected might be a historical item that has some fundamental characteristics associated with an area of interest.

8 Twenty

Design is about creating a desirable solution to a given scenario through a series of stages. A design can also be conducted or inspired through a reverse process, where an outcome is essentially understood at the outset but its potential application or meaning needs to be discovered.

In such situations, interaction through play can create an understanding of what a design might become, and assist in revealing its purpose.

The opportunity to design through a reverse process is not uncommon, since many factories produce vast amounts of waste material as an unfortunate by-product of a core outcome. Excess material is too often overlooked, but assigning it to a viable outcome can transform the waste into something that is also respected.

The unwanted units of surplus material in a factory have a particular characteristic that can be exploited in repurposing, in that they are identical, repeated units and they are created in vast numbers due to the mechanized processes. In such circumstances the design process is simply to identify an effective role for the surplus.

Taking a wood offcut that is approximately 100 x 50 x 25 mm, add two simple marks and repeat these marks on nineteen other identical offcuts. The marks might be simple cuts, holes, or both but all twenty offcuts must be exactly the same. Having created twenty identical objects, find a viable and different application for each of the offcuts. Originality is of the utmost importance, and the more creative the solution and the better the narrative associated with the outcome, the greater the likelihood of it being accepted.

9 What do they like?

Designing something for a particular audience requires an acute awareness of what it is that they appreciate or subscribe to. An audience is not necessarily going to be a particular gender, a specific age, or from a certain culture, and its members can initially appear to have nothing in common. Primary research, understanding, and observation will support the development of a user profile. A set of references that are expected to be in an area of interest for the potential audience can prompt thoughts, and may reaffirm or reject initial ideas. The creation of a profile is an indicator, a suggestion of what should be considered, but it needs to be conducted carefully and without bias or prejudice. A series of options, connected to a variety of themes, may emerge through the profiling stage, as it is unlikely that a single scenario will be appropriate for all, but there is likely to be a common thread.

Collect a series of secondary references based on the themes "I want to be different" and "personalization," prior to identifying a profile for three distinct individuals.

10 More tea please

The process of making tea varies considerably at both an international and local level. The ceremony of tea is important and the way that it is created, served, and enjoyed is significant. Tea that is made with haste, poured without due consideration and drunk from substandard cups is likely to be a less memorable experience than a tea that is respected and savored.

It is not simply the tea that is important, but also all the peripheral elements that contribute to the overall experience. The atmosphere or setting in which tea is made, and the individuals who it brings together can also differ. These varied experiences can often be observed in close proximity to each other.

Identify a public setting for drinking tea, such as a local café or restaurant, and experience the manner in which it is made, the quality of service and the different individuals who are attracted to the same venue. Having encountered the experience, select a different place the following day to drink tea and make the same observations. Changing the location for drinking tea every day will eventually steer you away from comfort zones, towards a previously unknown environment, where valuable information can be gained on how individuals interact, their expectations, their mannerisms, and their experiences.

11 Apple collecting kit

Understanding the needs of others is necessary when designing for others. It is important to know what habits, routines, and conventions are followed in doing something. Such characteristics might appear to be peculiar or bizarre, but the reasons behind why they are embraced need to be understood.

Different groups have different ways of doing things and it is not uncommon for children to imitate the activities of others, even if they are unable to comprehend the reasoning behind such actions. A particular activity for a child might have a completely different meaning to an adult.

Children often gravitate around a group culture, conducting activities in teams. The collecting of apples from an orchard is an activity that might be conducted by a group of children. Collecting apples is perhaps regarded by many as an enjoyable activity, but for the ten-year-old child it can be more than an activity — it can also become an adventure, especially when their imagination begins to make connections with associated practices.

Design a non-gender-specific apple collecting kit for children aged ten years old. What does the child need to collect apples and why is apple collecting so important? Do they want the apples for themselves or do they want to collect the apples for a relative or friend? Where will the apple collecting take place, and what dangers do they need to consider?

Things that children may want in their apple collecting kit might include:

- A box that they are able to stand on to reach the very best apples
- A sign to warn that harmful insects may be present during the collection process
- A series of small cones or tape to secure the apple collecting area
- An apple collector's safety uniform
- A variety of tools for the safe removal, peeling and cutting of apples at the scene
- An apple crusher and beaker to make a refreshing apple juice drink
- A tagging system to let other gangs know that the best apples have been collected
- A compass, map and notebook to locate and record the best parts of an orchard
- A basic emergency kit to treat stings and bites
- An apple collecting lunch kit and flask

It is unlikely that all these ideas could be incorporated into a single solution, and therefore priorities would need to be identified and additional ideas taken into account.

12 Roadside beauty

When a tyre needs changing on a vehicle it is not uncommon for there to be poor visibility at the time, or for it to occur during challenging weather conditions. There is never a good time for a tyre change. The process involves the use of frequently difficult-to-access tools, is dangerous, dirty, and generally unpleasant.

Design a roadside assistance kit for changing a tyre that will be produced by a leading skincare company who wants to focus on care for hands, nails, and hygiene.

Consideration should be given to the priorities of the product, but also to the emotional requirements of the user. The stigma of grime, along with the hassle and bother of changing a tyre, needs to be considered as an opportunity to refresh and cleanse. Analogous references might include dressing tables, manicure sets, and beauty salon workstations.

13 Need to rest

In Southeast Asia the act of squatting rather than resting on a chair is common, and it is possible for individuals to remain in such a position for long periods of time without any obvious discomfort.

Mixing cultural and historical references, design a seating solution for a Western culture that is specifically influenced by the practice of squatting in Southeast Asia.

Consider the pure and efficient outputs of Marcel Breuer and the effective use of tubular steel to create works such as the 1928 Cesca armchair. Following and understanding such principles, look to create a seating solution for squatting that attempts to echo the functionality of Marcel Breuer's works while introducing an Eastern practice to a broader Western audience.

14 My favourite color is...

An individual might often say what their favourite color is, but in the field of design, stating that a design should be a particular color is useless without a specific reference. There are many colors that could be assigned to a particular genre. A color can appear to be many things including fresh, dirty, bright, and cheap. Selecting a particular color, collect a hundred different examples of that color from a range of images, and then categorize them according to characteristics such as elegance, sophistication, fragility, and robustness.

15 Strip light

The creation of a simple-looking object is not necessarily as simple as it may seem if it has been created with careful consideration. Many versions will have been considered prior to that outcome being revealed. Simple designs frequently have problems or issues that need to be carefully resolved. Every aspect of a design needs to be explored from every possible angle in order to understand the most effective and desirable way to achieve the intended outcome.

Attempting to conclude a design that has not been fully thought through is likely to require continuous rethinks as the idea stumbles and stalls forward. Design is a tool that can remove problems through considered assessment of all options. It is a process where all possibilities need to be scrutinized, and where the investment of thought can often remove or lessen any pending complications.

Using thin, flexible card, create numerous strips that can be interwoven or even tangled to create a tangible form for a simple suspended lighting surround. Although tape might initially be used to hold the strips together, thought will need to be given to a more substantial form of fixing. Adhesives are often acceptable, but they can also be subject to failure, which would present difficulties if used in a final product. More robust fixing solutions can be considered but might be a distraction to the flowing lines that the paper strips create. Care needs to be taken in selecting the correct solution.

Having achieved an acceptable outcome in card, recreate the design using thin, flexible strips of ABS plastic, being sure not to interrupt the flowing lines that were initially created. As the lighting surround is to be suspended, thought must also be given to how this can be achieved in an aesthetically appealing manner.

The lighting surround is not the actual light, but it has the capacity to function in directing and controlling light. Consider all the various light sources that could be used and how any finish applied to the plastic might complement the design.

16 Recalling a memory

It is usually possible to recall something to memory, but it is often just a basic reference that is recalled and it can be difficult to remember the detail without a secondary trigger or reference.

A song, a memory, or an idea requires inspiration and reference material that can be developed and enhanced if something is to become real.

Singing on stage with no audience or backing band is as difficult as generating an idea successfully without physical or mental support.

Collect as much reference material as possible, but know what to use, when to use it, and how.

Conclusion

> "See and listen and see and listen again and again."
> **YeLi and Dave Bramston**

Think of a song and now sing it out loud. This is almost impossible to do except for a few lines or maybe a chorus. When there is accompanying music it is a little bit easier, but listen to the enthusiastic sounds emanating from the audience at a concert and you will soon realize that even with professional acoustic support it can take a long time to actually hear an accurate rendition.

The analogy is much the same in the design process, in the sense that it is often possible to see things in the mind, but quite a different thing to try and effectively communicate them. There is a need to be surrounded by visual stimulation, images that can assist and prompt thinking even if they are not directly related to the matter in hand.

To be creative without visual input is not impossible, but it can be more difficult. The hoarding of objects to rouse and encourage thinking is a natural tendency for a designer. It is equally important to get out of the studio environment and get involved, experience things and communicate with others, to see what is actually going on. There should never be a situation where the design process becomes stagnant or uninteresting, and if such times do occur there is a need to consider things from an alternative perspective and to see what emerges.

Isolation in the design process may be needed to think for a short time, but questions should be looked at in unison, rather than a single individual trying to work alone. Almost without exception the design process is a team effort, even if there is a recognized maestro steering the process.

The research work for authoring this book did not start when the publication was conceived, as that was only the identification of a collection of experiences to be drawn together. Unknowingly, the research began decades before, on multiple levels, and by many individuals.

Glossary

Adlib
Adlibbing is a process often adopted in the initial stages of idea generation. It is a spontaneous action or the result of an unexpected outcome.

Aesthetic
Aesthetics are the emotional characteristics and considered design values associated with the overall perception of an object.

Analogous
Two or more different or seemingly unrelated objects are analogous when they have a particular characteristic, feature, or use in common.

Animate
Animate refers to being alive or having life.

Anthropologist
The anthropologist observes and develops an understanding of the fundamental mannerisms, behavior, attitudes, social beliefs, and customs of societies and cultures.

Artifacts
An artifact is an object that is made by a person.

Banal
Something that is banal lacks originality and is uninspiring. A banal object is something that might be considered common or dull.

Benchmark
A benchmark is a particular standard or reference that can be used to gauge other similar things by.

Bespoke
A bespoke item is unique and created to individual criteria.

Blue-sky thinking
Blue-sky thinking refers to high creativity that does not need to connect with conventional understanding and encourages the imagination to explore without constraints.

Brainstorming
Brainstorming is an activity typically conducted in small groups to elicit, explore, and share initial thoughts. A facilitator usually directs the sessions, which can take place throughout the design process, although such sessions are particularly common around the initial outset of the design process. A brainstorming session can adopt many formats.

Catalyst
A catalyst is something that speeds a process up and enables outcomes to occur faster. A verbal or visual catalyst is something that triggers new creative ideas and thoughts.

Chemostat
A chemostat is apparatus used by biologists to observe and gauge the activity of specific bacterial cultures under different conditions. The development of cultures can be controlled through the monitoring of parameters such as nutrient availability or flow rate.

Comfort zone
A comfort zone is a limited area or range of experience with which an individual is familiar and unchallenged. Remaining in one's comfort zone prevents individuals from experiencing the unknown. It is important to explore beyond the comfort zone in order to discover new and original experiences.

Creativity rut
A creativity rut is a situation in which the imagination is stuck and an individual has ceased to make progress. This may be caused by overthinking a particular idea, or by insufficient initial inspiration.

Cross-disciplinary
Something is cross-disciplinary if it makes reference to two or more unrelated subject disciplines.

Eclectic
Eclectic refers to when a broad range of unrelated styles are brought together to create inspired outcomes.

Hackneyed
A hackneyed idea is one that has been overused and is now dull and uninspiring.

Hybrid
When an outcome is produced from two or more different concepts or disciplines, it is a hybrid.

Inanimate
An inanimate object is one that is not alive and shows no sign of having life.

Info dump
An info dump is an informal method for presenting and discussing a range of visual sources that could be inspirational in the development of an original product.

Interdisciplinary
Something is interdisciplinary if it makes reference to two or more related subject disciplines, for instance, various subject disciplines within the arts.

Junk model
A junk model is a simplified model of a product, created using readily available materials, which can help to visualize an idea in its early stages.

Lateral thinking
Lateral thinking explores unexpected and alternative approaches to problem solving. This abstract approach provides an opportunity to create original outcomes.

Literal thinking
Literal thinking follows familiar routes and uses unoriginal sources in problem solving.

Mental baggage
Mental baggage refers to an individual's preconceived ideas on a particular topic, which can prevent or significantly hinder original thoughts.

Mind-set
Mind-set refers to an individual's distinct opinions and ideas, which can be difficult to overcome if it becomes too fixed or familiar.

Open mind
Someone with an open mind is receptive to new ideas, views, and arguments.

Primary research
Primary research is the research that you carry out firsthand, rather than by referring to other published sources.

Profiles
User profiles are a means of summarising the interests and requirements of a particular audience. Understanding the products that a target group already uses can provide insight when developing new products.

Glossary

Prototype
A prototype is the first version produced of a designed artifact. It offers an opportunity for evaluation or final sign-off of a product.

Role play
Role play is the acting out of a particular idea or problem, with limited or no resources. Role play can help you to identify issues or opportunities at the beginning of the design process.

Scenario
A scenario is a setting for a particular activity or problem. Recognizing a problem scenario can trigger creative thought to overcome the problem.

Secondary research
Secondary research involves gathering information that has already been sourced or published by a third party. This can work well in conjunction with primary research.

Stereotype
An overly generic and simplistic perception of something. A stereotype is a generalization about a particular category or group, which is often inaccurate and misleading.

Up-cycling
Up-cycling refers to the process of reusing or reworking something that was previously abandoned or disposed of, while increasing its value or usability.

Bibliography

Books

Aldersey-Williams, H.
King and Miranda: The Poetry of the Machine
(Blueprint monographs)
Fourth Estate 1991

Alessi, A.
The Dream Factory: Alessi Since 1921
Konemann UK Ltd 1998

Antonelli, P.
Humble Masterpieces: 100 Everyday Marvels of Design
Thames & Hudson 2006

Antonelli, P.
Mutant Materials in Contemporary Design
Museum of Modern Art 2005

Bakker, G. and Ramekers, R.
Droog Design—Spirit of the Nineties
010 Uitgeverij 1998

Benyus, J.M.
Biomimicry
William Morrow 1997

Bloemendaal, L.
Humanual
Uitgeverij BIS
Amsterdam 2002

Börnsen-Holtmann, N.
Italian Design
Benedikt Taschen 1994

Bouroullec, R. and E. Bouroullec
Bivouac
Centre Pompidou-Metz 2012

Brownell, B.
Transmaterial
Princetown Architectural Press 2006

Buckminster Fuller, R.
Operating Manual for Spaceship Earth
Southern Illinois University Press 1969

Campana, F. and H. Campana
Campana Brothers: Complete Works (So Far)
Rizzoli International Publications 2010

Coates, N.
Guide to Ecstacity
Laurence King Publishing 2003

Dempsey, A.
Styles, Schools and Movements
Thames & Hudson 2004

De Noblet, J.
Industrial Design: Reflection of a Century
Flammarion 1993

Dixon, P.
Futurewise: Six Faces of Global Change
Harper Collins 1998

Dixon, T. et al.
And Fork: 100 Designers, 10 Curators, 10 Good Designs
Phaidon Press 2007

Fiell, C. and P. Fiell
Design for the 21st Century
Taschen 2003

Fleck, R. and R. Fuchs
Erwin Wurm
Hatje Cantz, Germany 2006

Forty, A.
Objects of Desire
Thames & Hudson 1986

Fuad-Luke, A.
The Eco-Design Handbook
Thames & Hudson 2002

Fukasawa, N.
Naoto Fukasawa
Phaidon Press 2007

Fulton Suri, J. and IDEO
Thoughtless Acts?
Chronicle Books 2005

Gamper, M.
100 Chairs in 100 Days and its 100 Ways
Dent-De-Leone 2007

Bibliography

Gershenfeld, N.
When Things Start to Think
Hodder & Stoughton 1999

Hauffe, T.
Design: A Concise History
Laurence King Publishing 1998

Jansen, T.
Theo Jansen: The Great Pretender
00 Uitgeverij: 1st edition 2007

Jensen, R.
The Dream Society
McGraw-Hill 1999

Kaku, M.
Visions
Oxford University Press 1998

Kelley, T.
The Ten Faces of Innovation
Doubleday 2005

Kung, M.
Freitag: Individual Recycled Freeway Bags
Lars Muller Publishers 2001

Lupton, E.
Skin
Laurence King Publishing 2002

MacCarthy, F.
British Design since 1880
Lund Humphries 1982

Meneguzzo, M.
Philippe Starck Distordre
Electa/Alessi 1996

Moors, A.
Simply Droog
Droog Design, revised edition 2006

Museum of Design, Zurich
Freitag: Out of the Bag
Lars Muller Publishers 2012

Naylor, N. and R. Ball
Form Follows Idea: An Introduction to Design Poetics
Black Dog Publishing 2005

Nendo
Nendo: 10/10
Die Gestalten Verlag 2013

Nendo, O.S.
Nendo
Daab; Mul edition 2008

Newman, M.
Wurm Erwin
Photographers' Gallery 2000

Norman, D.A.
The Design of Everyday Things
MIT Press, 2nd revised and expanded edition 2013

Papanek, V.
The Green Imperative
Thames & Hudson 1995

Pink, D.
A Whole New Mind
Cyan 2005

Smith, P.
You Can Find Inspiration in Everything
Violette editions 2001

Sozzani, F.
Kartell
Skira Editore Milan 2003

Sweet, F.
Frog: Form Follows Emotion
Thames & Hudson 1999

Thompson, D.
On Growth and Form
Cambridge University Press 1961

Walker, S.
Sustainable by Design—Explorations in Theory & Practice
Earthscan Ltd 2006

Weschler, L. and T. Jansen
Strandbeest: The Dream Machines of Theo Jansen
Taschen Gmbh 2014

Wilcox, D.
Variations on Normal
Kindle edition

Wurm, E.
Erwin Wurm
DuMont Literatur und Kunst Verlag GmbH & Co KG 2009

Journals and magazines

Abitare
Artform
AZURE
Blueprint
b0x
DEdiCate
Design
Design Week
domus
dwell
Egg
FRAME
frieze
FRUiTS
icon
ID
INNOVATION
intramuros
Kult
Lowdown
MARK
Metropolis magazine
mix
MODO
MONUMENT
newdesign
Product Design WORLD
Stuff
surface
T3
TWILL
vanidad
W magazine
wallpaper

Blogs

booooooom.com
coolhunting.com
core77.com
designboom.com
design-milk.com
dezeen.com
formfiftyfive.com
itsnicethat.com
swiss-miss.com

Contacts

alessi.com
adamverity.co.uk
astrostudios.com
barberosgerby.com
biocouture.co.uk
bouroullec.com
campanas.com.br
chen-min.com
c-lab.co.uk
danielrybakken.com
diegostocco.com
dominicwilcox.com
droog.com
erwinwurm.at
freitag.ch
frogdesign.com
ideo.com
jannishuelsen.com
j-me.com
kraud.de
lala-lab.com
lukejerram.com
markzirpel.com
mars-hwasung.com
matthias-studio.com
nendo.jp
oliver-schick.com
paulcocksedgestudio.com
paulogoldstein.com
philips.co.uk
remyveenhuizen.nl
stefanogiovannoni.it
strandbeest.com
stuartwalker.org.uk

studioball.co.uk
studiolibertiny.com
studio-stephanschulz.com
tinaroeder.com
tuckerviemeister.com
vibrazioniartdesign.com
whitecube.com
wiekisomers.com
xavierhufkens.com
yebramston.com

Index

99p orchestra 159
100 Chairs in 100 Days 46, 128

absorb, imagination 10–18
Acetobacter bacteria 94
Acetobacter xylinum 97
added values 115–22
adlibs 18
adopting responsibility *see* responsibility adoption
adoration 114, 130
aesthetics 30
Alessi, Alberto 80–1
Altar of Things 60–1
analogous objects 32, 36–7
Anatomy Series 22–3
anecdotal references 24–8
animate objects 76
anthropologists 79
Anti-Design groups 91, 100
apple collecting kits 164
Archigram 91
Archive Series 23
Archizoom 91
Arcophonico 48
aromas 112
artifacts 16
assumptions 66–8
Astro Studios 41, 74
audience considerations 163
audio 84–5, 113
Avio Chair 143

Bakker, Gijs 44, 82
Ball, Ralph 22–3, 54–5
Bamboo Blue XL 128, 129
Barber & Osgerby 88–9
Beijing Bowl 107–9
benchmarks 53
Berg, Dylan van den 10–14
Berlino Bench 73
bespoke objects 51
bioactive materials 92–7
Biocouture collections 94–5
Bio-light 92–3
blogs 15
Blue sky thinking 151–5
Boland, Howard 93–4
Bourgie table lamp 66–7
Bouroullec, Ronan and Erwan 37
brainstorming 38–41
Bramston, Dave 169
Breuer, Marcel 166

Buddha Maitreya sculpture 110

Campana brothers 118–21
Candy collection 118–20
Cass, Garry 94
cellphones 131
ceramics 30–1
ceremony, performance 60–1
chairs: 100 Chairs in 100 Days 46, 128; Anatomy Series 22–3; Bouroullec 37; Concrete Furniture 98–9; Fehling & Peiz 73; Gamper & Spehl 73, 128; Goldstein 139; Hayon Studio 128; hidden seating 157; improvisation 42–4, 46; meetings 161; Plastic Gold 54–5; Roeder 52; Schick 78, 128; transformation 52, 54–5; Transparent Chair 56–7; Vibrazioni Art Design 143; Vladimir 72, *see also* stools
Chairstoolbench 73
challenging understanding 70–3
character adoption 75–8
chemostats 92–4
Chen, Kelly 59
Chen, Min 114–18
Chest of Drawers 44–5
The Chinese Hat 30–1
Chinese Stools 10, 14
Chow, Cat 53
Ciniti, Laura 93–4
C-Lab 93–4
Clerkin, Carl 77
collecting kits 164
color 107–9, 166
comfort zone 91
communication 102–3
Concrete Furniture 98–9
conflicts 124–31
contrasts 88–92
control, perceptions 125–9
creativity ruts 21
cross-disciplinary references 15, 158
Cruz, Tania da 75
cultural references 79–81
Custom Built Orchestra 47–9, 51
"Custom" or "Hacked" cellphone 131

Dassasso, Alberto 143

Daylight Comes Sideways 121–2
daylight effects 121–3
developing the idea 154–5
Droog collection 44–5
Droog Design 82–3

eclectic references 14, 158
Edison, Thomas A. 91
Eigruob table lamp 66–7
Einstein, Albert 20
elegance 158
emotional responses 114
environment 21
Eric Morel Design 75
Erkök, Mehmet 131
Escherichia coli (E. coli) 97
evolving reality 151–5
Experibass 48
experiences 9–13
experimentation/experimental beauty 87–103; communication 102–3; exploration 92–101; materials 88–97
Experiviolin 48
exploration 92–101

F13 Top Cat messenger bag 144–5
Family Vase collection 82–3
Fassi, Davide 56
Fat Car Convertible 124–5
favorite colors 166
Fehling, Yvonne 73, 77
feigned standpoints 20
Fire lamp 36
Fishingstick 146–7
"form follows function" 91
Found light 140–1
Franklin, Donna 94
Fratesi, Enrico 126–7
Freitag brothers 144–5
Fuller, Buckminster 29, 88

Gamper, Martino 46, 73, 128
Garci-Crespo, Lucia Lopez 146–7
garments 94–5
Gijs Bakker Design 44
Giovannoni Design 80–1
Goldstein, Paulo 138–9
"gong" 114–15, 118
Green Chicken Rocking Horse 100

Index

hackneyed standpoints 20
harmony 160
HAT lamp 31
Hatoum, Mona 71
Hayon, Jamie 100–1
Hayon Studio 128
Hello concept 161
hidden seating 157
Hide 'n Seek 102–3
HIFA (Hubei Institute of Fine Arts) 33
historical references 82
honesty 140–5
Hubei Institute of Fine Arts (HIFA) 33
Huelsen, Jannis 96–7
hybrid instruments 48

IDEO team 40–1, 43, 97
illuminated beauty 156
imagination 9–61; absorption 10–18; observation 42–61; thoughts 19–41
ImaginationLancaster Design Research Center 136–7
improvisation 42–9
inanimate objects 76
info dumps 16–17, 159
inner-city street seating 42
inspiration 42–3, 56–9, 148–9
interdisciplinary references 15, 158
Internet 21

Jansen, Theo 148–9
Jerram, Luke 28
j-me toothbrush holders 76
journals 15, 158
junk modeling 43

Klunder, Nienke 101

"laboratory" approach 90
LaLa Lab 36
lamps *see* lighting
language 29–35
Lanterne Marine vase 88–9
lateral thinking 18
Lather lamp 136–7
Lee, Suzanne 94–5
Leirner, Jac 70
Libertiny, Tomáš 53
lighting: Bio-light 92–3; Campana brothers 118–19; C-Lab 93–4;

daylight effects 121–3; Goldstein 138; illuminated beauty 156; Jac Leirner 70; LaLa Lab 36; Mars Hwasung Yoo 31; Mona Hatoum 71; Ms. Kitty 107–9; Oki Sato 66–7; plastic containers 141–2; strip lights 167; Studio Wieki Somers 10–14; Tsinghua International Foundation 134–5; Walker 136–7; Yuxi 140–2
literal thinking 64
Little Crawly Thing 77
Little Light 3 70
Littler, David 32
Li, Ye 169
Lollipop table lamp 118–19
Luminopiano 48
Lunar 39–41, 56
Luxury Skimming Stones 55
Luya, Shao & others 134–5

McCarthy, Laura 15
Magna 126
Mama, Jack 92–3
Mao, Natalie 30
materials 88–97
meeting places 69, 161
memory recall 168
Memphis group 91
mental baggage 19–23
mental notes 19
mental role play 64
messenger bags 144–5
Micro "be" collection 94
Microbial Home Probe, Bio-light 92–3
microbial materials 92–7
Mikkelsen, Maria Kirk 90
mind-sets 33
Mitate lamp collection 10–13
Moggridge, Bill 10, 43
Morel, Eric 75
Mr Chin 80
Ms. Kitty 107–9
"Mu" 115–18
multiple object creation 162
Murano glass vases 88–9
Museman 84–5
music 47–51, 159

Nam, Miyoung 125, 126, 127

National Palace Museum, Taiwan 80–1
Naylor, Maxine 22–3, 54–5
Nendo 56–7, 66–7
New Original collection 82–3
notebooks 16–18

Oblivious iPhone User 102
observation 42–61
odors 112
Off-Cut collection 46
Offset Vertical Cut 22
openness 138–9
orchestras 47–9, 51
OrienTales 80–1

packaging 155
Paul Cocksedge Studio 20
Peiz, Jennie 73, 77
perceptions, sensory 105–31
performance, ceremony 60–1
personal experiences 9–13
Philips Design Probe 92–3
physical role play 64
The Pink Ceramic Ballet Shoes 15
plastic bowls 107–9
plastic containers 141–2, 148–9
Plastic Gold 54–5
Platinum, Pliessnig 3
play 100–1
"Play Me, I'm Yours" concept 28
Playmobilia stools 75
Pliessnig, Matthias 3, 16–19
polypropylene packaging straps 146–7
Power Ranger Kimono trading cards 53
primary research 14
prior experience 10–13
profiles 74
project briefs 156–68
prototyping 43
purpose 65

Quilt chair 37

Rachev, Vladimir 72
Rag Chair 44
Ramakers, Renny 82
Ratzlaff, Jörg 84–5
reality evolution 151–5
recall 168
refreshments 163

Remy, Tejo 44–5, 98–9
Repaired Anglepoise 1 work 138
The Repaired Director's Chair 139
research: primary research 14; secondary research 74
resources: inspiration 148–9; sustainability 137
respect 130–1
responsibility adoption 132–49; inspiration 148–9; sustainability 134–7; tinkering 138–47
rest periods 166
roadside beauty 165
Rocking Sausage Hot Dog 100–1
Roeder, Tina 52
role play 64
Ronan and Erwan Bouroullec 37
Rybakken, Daniel 121–3

samplers 32
Sato, Oki 56–7, 66–7
scale 102–3
scenarios 29
Schick, Oliver 78, 128
Schulz, Stephan 60–1
scrapbooks 16–17
seating *see* chairs; stools
secondary research 74
senses, engagement 106
sensory perceptions 105–31; added values 115–22; conflicts 124–31; control 125–9; unexpected 115–22
Shaw, George Bernard 38
Showtime Armchair BD 128
simplicity 146–7
sketchbooks 16–17
smell 112
Soft Concrete Bench 99
Somers, Wieki 10–14
Sony Walkman II 84–5
Sottsass, Ettore 91
sound 84–5, 113
Spehl, Rainer 73
Sphere chandelier 118–19
stereotypes 68

Still Lives works 77
Stocco, Diego 47–9, 51
stools: Chairstoolbench 73; Guan Ziyin 24–8; Playmobilia stools 75; Studio Wieki Somers 10, 14; Xylinum stool 96–7
Strandbeest 148–9
Street chair, Nanchang Street, China 42
street seating 14
Stress-o-stat 93–4
strip lights 167
Studio Droog 82–3
Studio Libertiny 53
Studio Recordings 53
Studio Wieki Somers 10–14
Stuhlhockerbank 73
Subconscious Effect of Daylight 121–2
subtlety 124–5
Sugar XL 128, 129
sun-bleached plastic bowls 107–9
SUPERSTUDIO 91
Surface Daylight 1 & 2 121–3
Sushi series 118–19
sustainability 134–7
Swaminathan, Miep 93

taboos 91
Tattoo table 120–1
tea, refreshments 163
tension 160
terminology, language 32–4
Textural Flute 49
texture 110–11
themes 75–85
thinking differently 63–85; profiles 74; themes 75–85; understanding 64–73
thoughts 19–41
Tonal Metals 48
touch 110–11
trading cards 53
transformation 52–5
Transparent Chair 56–7
Tsinghua International Foundation program 134–5

Tsinghua University, Beijing, China 43
Tune 'n Radio 153

Undercurrent 71
understanding: challenging 70–3; thinking differently 64–73
unexpected: added values 115–22; elegance 158; sensory perceptions 115–22
Unfoldable Chair 72
Unger, Götz 40

vases 82–3, 88–9
Veenhuizen, René 98–9
Verity, Adam 65, 68
Vibrazioni Art Design 143
Viemeister, Tucker 69

Walker chair, Schick 78, 128
Walker, Stuart 136–7
Walkmans 84–5
Watch Sculptures works 102–3
Water Organ, Zirpel 50–1
Watt? 20
Weldgown 53
White Billion Chairs 52
Wilcox, Dominic 55, 102–3
Wilde, Oscar 39
Wire light 136–7
workspace considerations 160
Wouter Geense Design Studio 153
Wurm, Erwin 124–5

Xiao-hua, Mao 30–1
Xylinum stool 96–7

Yellow Soldier light 141–2
Yoghurt Pot Chandelier 134–5
Yoo, Mars Hwasung 30–1
Yuxi, Zhang 140–2

Zelkowitz, Hy 38
Zhou Hui 146–7
Zirpel, Mark 50–1
Ziyin, Guan 24–8

Credits

Front cover
Peng Tun

Page 3 fig.0.1
Ieva Saudargaité

Page 11 fig.1.1
Fabrice Gousset

Page 12 fig.1.2
Fabrice Gousset

Page 13 fig.1.3
Fabrice Gousset

Page 14 fig.1.4
Pien Spijkers

Page 15 fig.1.5
Laura McCarthy

Page 17 fig.1.6a and fig.1.6b
Matthias Pliessnig

Page 18 fig.1.6c
Matthias Pliessnig

Page 19 fig.1.7
Matthias Pliessnig

Page 19 fig.1.8
Matthias Pliessnig

Page 20 fig.1.9a and fig.1.9b
Richard Brine

Page 22 fig.1.10
Ralph Ball and Maxine Naylor

Page 23 fig.1.11a–fig.1.11d
Ralph Ball and Maxine Naylor

Page 24 fig.1.12
Guan Ziyin

Page 25 fig.1.13
Guan Ziyin

Page 26 fig.1.14
Guan Ziyin

Page 27 fig.1.15
Guan Ziyin

Page 28 fig.1.16
Luke Jerram

Page 30 fig.1.17
Natalie Mao

Page 31 fig.1.18
Stephanie Wiegner

Page 33 fig.1.19
HIFA

Page 36 fig.1.20
Courtesy of Yuki IIDA, Lala Lab

Page 37 fig.1.21
Ronan and Erwan Bouroullec

Page 40 fig.1.22
IDEO

Page 41 fig.1.23
Courtesy of Astro Studios

Page 42 fig.1.24
Dave Bramston

Page 43 fig.1.25a and fig.1.25b
Tsinghua University

Page 44 fig.1.26
Gerard van Hees

Page 45 fig.1.27
Bob Goedewagen

Page 47 fig.1.28
Gianfilippo de Rossi

Page 49 fig.1.29
Gianfilippo de Rossi

Page 50 fig.1.30a and fig.1.30b
Mark Zirpel

Page 52 fig.1.31
Guido Mieth

Page 53 fig.1.32
René van der Hulst

Page 54 fig.1.33
Ralph Ball and Maxine Naylor

Page 55 fig.1.34a and fig.1.34b
Dominic Wilcox

Page 57 fig.1.35
Masayuki Hayashi

Page 59 fig.1.36
Kelly Chen

Page 61 fig.1.37a–fig.1.37d
Matthias Ritzmann

Page 65 fig.2.1
Adam Verity

Page 66 fig.2.2a
Akihiro Yoshida

Page 67 fig.2.2b
Akihiro Yoshida

Page 68 fig.2.3
Adam Verity

Page 69 fig.2.4
Courtesy of Tucker Viemeister

Page 70 fig.2.5
Courtesy of Galeria Fortes Vilaça and White Cube

Page 71 fig.2.6
Murat Germen
Courtesy of Arter, Istanbul and White Cube

Page 72 fig.2.7
Vladimir Rachev

Page 73 fig.2.8
Yvonne Fehling and Jennie Peiz

Page 74 fig.2.9
Courtesy of Astro Studios

Page 75 fig.2.10
Manuel Rio Casali

Page 76 fig.2.11a and fig.2.11b
j-me

page 77 fig.2.12
Frederik Busch

Page 78 fig.2.13
Michael Himpel

Page 80 fig.2.14
Giovannoni Design

Page 81 fig.2.15
Giovannoni Design

Page 82 fig.2.16
Mo Schalkx

Page 83 fig.2.17
Mo Schalkx

Page 85 fig.2.18
V. Goico (Image Museman)

Page 89 fig.3.1
Edward Barber and Jay Osgerby

Page 92 fig.3.2
Philips Design

Page 94 fig.3.3
C-Lab

Page 95 fig.3.4
Science Museum / Science & Society Picture Library

Credits

Page 96 fig.3.5
Jannis Huelsen

Page 97 fig.3.6
Jannis Huelsen

Page 98 fig.3.7
Ernst Moritz

Page 99 fig.3.8
Ernst Moritz

Page 101 fig.3.9
Jamie Hayon and Nienke Klunder

Page 102 fig.3.10
Dominic Wilcox

Page 103 fig.3.11
Dominic Wilcox

Page 107 fig.4.1
Qicong Lin

Page 108 fig.4.2
Dave Bramston

Page 109 fig.4.3
Qicong Lin

Page 110 fig.4.4a and fig.4.4b
Dave Bramston

Page 111 fig.4.5
Dave Bramston

Page 112 fig.4.6
Dave Bramston

Page 114 fig.4.7a and fig.4.7b
Min Chen

Page 115 fig.4.8
Min Chen

Page 117 fig.4.9a–fig.4.9c
Min Chen

Page 119 fig.4.10a and fig.4.10b
Hana Klimova

Page 121 fig.4.11
Daniel Rybakken

Page 121 fig.4.12
Daniel Rybakken

Page 123 fig.4.13
Kalle Sanner

Page 124 fig.4.14
Erwin Wurm and Xavier Hufkens

Page 127 fig.4.15
Dave Bramston

Page 129 fig.4.16
Y&Bramston

Page 131 fig.4.17
Mehmet Erkök

Page 134 fig.5.1
Qicong Lin

Page 137 fig.5.2
Stuart Walker

Page 137 fig.5.3
Stuart Walker

Page 138 fig.5.4
Paulo Goldstein

Page 139 fig.5.5
Paulo Goldstein

Page 140 fig.5.6
Zhang Yuxi

Page 141 fig.5.7
Zhang Yuxi

Page 143 fig.5.8
Alberto Dassasso

Page 143 fig.5.9
Callo Albanese

Page 145 fig.5.10
Peter Würmli

Page 147 fig.5.11
Dave Bramston

Page 148 fig.5.12
Theo Jansen

Page 149 fig.5.13
Theo Jansen

Page 153 fig.6.1
Wouter Geense Design Studio

Acknowledgments

Thank you to all the designers, artists, photographers, and researchers who have supported this project and provided exciting images and statements.

The information has been sourced from all over the world and has involved young designers along with influential leaders within their respective disciplines. The involvement of all of these individuals is really appreciated.

Many thanks also to the programs where design workshops supported the research for the book, including the MA International Design Enterprise program at the University of Lincoln, UK, the Chinese University of Mining and Technology School of Arts & Design, Xuzhou, China, the Industrial Design program at Hubei Institute of Fine Arts, Wuhan, China, the Industrial Design program and the International Foundation program at Tsinghua University, Beijing, China, the Industrial Design program at Philadelphia University, USA, the International Foundation course students at the China Academy of Fine Arts (CAFA IFC), Beijing, China, the design programs at the Guangzhou Institute of Technology, Guangzhou, China, the Sorrell Foundation National Art & Design Saturday Club, UK, and the Industrial Design program at Jiangnan University Wuxi, China.

Thanks also to the Guangzhou Valuda Group Co. Ltd, the Guangzhou Low Carbon Industries Association, and the British Council China who continually support grassroots design and low carbon practices.

Special thanks to Kate Duffy, Georgia Kennedy, and Felicity Cummins who provided the necessary and valuable support at Fairchild Books (UK) c/o Bloomsbury Publishing, London, UK.